Holocene Evolution of the Western Louisiana–Texas Coast, USA: Response to Sea-Level Rise and Climate Change

by

John B. Anderson
Department of Earth, Environmental and Planetary Sciences
Rice University, Houston, Texas 77251, USA

Davin J. Wallace
School of Ocean Science and Engineering, University of Southern Mississippi
Stennis Space Center, Mississippi 39529, USA

Antonio B. Rodriguez
Institute of Marine Sciences, University of North Carolina at Chapel Hill
Morehead City, North Carolina 28557, USA

Alexander R. Simms
Department of Earth Science, University of California
Santa Barbara, California 93106, USA

Kristy T. Milliken
Department of Earth, Environmental and Planetary Sciences
Rice University, Houston, Texas 77251, USA

THE
GEOLOGICAL
SOCIETY
OF AMERICA®

Memoir 221

3300 Penrose Place, P.O. Box 9140 ▪ Boulder, Colorado 80301-9140, USA

2022

Copyright © 2022, The Geological Society of America (GSA), Inc. All rights reserved. Copyright is not claimed on content prepared wholly by U.S. government employees within the scope of their employment. Individual scientists are hereby granted permission, without fees or further requests to GSA, to use a single figure, a single table, and/or a brief paragraph of text in other subsequent works and to make unlimited photocopies of items in this volume for noncommercial use in classrooms to further education and science. Permission is also granted to authors to post the abstracts only of their articles on their own or their organization's Web site providing that the posting cites the GSA publication in which the material appears and the citation includes the address line: "Geological Society of America, P.O. Box 9140, Boulder, CO 80301-9140 USA (https://www.geosociety.org)," and also providing that the abstract as posted is identical to that which appears in the GSA publication. In addition, an author has the right to use his or her article or a portion of the article in a thesis or dissertation without requesting permission from GSA, provided that the bibliographic citation and the GSA copyright credit line are given on the appropriate pages. For any other form of capture, reproduction, and/or distribution of any item in this volume by any means, contact Permissions, GSA, 3300 Penrose Place, P.O. Box 9140, Boulder, Colorado 80301-9140, USA; fax +1-303-357-1070; editing@geosociety.org. GSA provides this and other forums for the presentation of diverse opinions and positions by scientists worldwide, regardless of their race, citizenship, gender, religion, sexual orientation, or political viewpoint. Opinions presented in this publication do not reflect official positions of the Society.

Published by The Geological Society of America, Inc.
3300 Penrose Place, P.O. Box 9140, Boulder, Colorado 80301-9140, USA
www.geosociety.org

Printed in U.S.A.

GSA Books Science Editors: Joan Florsheim, Christian Koeberl, and Nancy Riggs

Library of Congress Cataloging-in-Publication Data

Names: Anderson, John B., 1944– author. | Wallace, Davin J., author. | Rodriguez, Antonio B., author. | Simms, Alexander R., author. | Milliken, Kristy T., author.
Title: Holocene evolution of the western Louisiana-Texas Coast, USA : response to sea-level rise and climate change / by John B. Anderson, Davin J. Wallace, Antonio B. Rodriguez, Alexander R. Simms, Kristy T. Milliken.
Description: Boulder, Colorado : The Geological Society of America, 2022. | Series: Memoir ; 221 | Includes bibliographical references. | Summary: "The Western Louisiana and Texas coast is vulnerable to sea-level rise due to low gradients, high subsidence, and depleted sediment supply. This Memoir describes the response of coastal environments to variable rates of sea-level rise and sediment supply during Holocene to modern time. This volume is a wake-up call about the potential magnitude of coastal change over decadal to centennial time scales"— Provided by publisher.
Identifiers: LCCN 2022016113 (print) | LCCN 2022016114 (ebook) | ISBN 9780813712215 (hardcover) | ISBN 9780813782218 (ebook)
Subjects: LCSH: Coast changes—Louisiana—Gulf Coast. | Coast changes—Texas—Gulf Coast. | Sea level—Louisiana—Gulf Coast. | Sea level—Texas—Gulf Coast. | Climatic changes—Louisiana—Gulf Coast. | Climatic changes—Texas—Gulf Coast. | Geology, Stratigraphic—Holocene. | Paleogeography—Holocene.
Classification: LCC GB459.25 .A44 2022 (print) | LCC GB459.25 (ebook) | DDC 551.45/70976—dc23/eng20220628
LC record available at https://lccn.loc.gov/2022016113
LC ebook record available at https://lccn.loc.gov/2022016114

Cover: (*Top*) Digital Elevation Model based on LIDAR data from Galveston Island, Texas, USA (data from https://coast.noaa.gov/). (*Bottom*) Simplified cross section through Galveston Island, Texas. This is a modified version of Figure 24 (p. 28 of this Memoir). Please refer to it for additional information.

Contents

ABSTRACT 1

INTRODUCTION 2

STUDY AREA 4

Sea-Level History 7

DATA SETS AND METHODS 9

Radiocarbon Dating 9

RESULTS 9

Late Pleistocene Coastal Evolution 9

MIS 5-2 Sea-Level Fall and Lowstand 10

MIS 2-1 Transgression 11

Sand Banks 12

Offshore Trinity and Sabine Incised River Valleys 13

Coastal Trinity, Sabine, and Calcasieu Valleys 15

Western Louisiana Chenier Plain 22

East Texas Barrier System 23

The Brazos Colorado Fluvial–Deltaic Systems 31

The Central Texas Coast 37

The South Texas Coast 52

DISCUSSION 59

Coastal Response to Climate Change and Decreasing Sea-Level Rise 59

Controls on Coastal Evolution 64

Antecedent Topography 64

Sea-Level Rise 65

Sediment Supply and Dispersal 66

Oceanographic Influences and the Texas Mud Blanket 69

Impacts from Severe Storms 69

Anthropogenic Impacts 70

Ongoing and Future Change 71

Wetland and Bay Impacts 71

Coastal Barriers 72

CONCLUSIONS 73

ACKNOWLEDGMENTS 74

REFERENCES CITED 74

The Geological Society of America
Memoir 221

Holocene Evolution of the Western Louisiana–Texas Coast, USA: Response to Sea-Level Rise and Climate Change

John B. Anderson
Department of Earth, Environmental and Planetary Sciences, Rice University, Houston, Texas 77251, USA

Davin J. Wallace
School of Ocean Science and Engineering, University of Southern Mississippi, Stennis Space Center, Mississippi 39529, USA

Antonio B. Rodriguez
Institute of Marine Sciences, University of North Carolina at Chapel Hill, Morehead City, North Carolina 28557, USA

Alexander R. Simms
Department of Earth Science, University of California, Santa Barbara, California 93106, USA

Kristy T. Milliken[*]
Department of Earth, Environmental and Planetary Sciences, Rice University, Houston, Texas 77251, USA

ABSTRACT

An extensive grid of high-resolution seismic data, hundreds of sediment cores, and a robust radiocarbon-age data set acquired over nearly four decades allows detailed analysis of Holocene coastal evolution of western Louisiana and Texas, USA. Results from this study provide a framework for assessing the response of a myriad of coastal environments to climate change and variable sea-level rise. Climate varies across the region today, spanning four climate zones from humid to semi-arid, and has fluctuated during the Holocene. The most notable changes were alterations between cool/wet and warm/dry conditions. Sea-level records for the northwestern Gulf of Mexico indicate an average rate of rise during the early Holocene of 4.2 mm/yr, punctuated by rates exceeding 10.0 mm/yr. After ca. 7.0 ka, the rate of rise slowed, and by ca. 4.0 ka, the average rate decreased from 0.6 mm/yr to 0.3 mm/yr. The current rate of sea-level rise in the region is 3.0 mm/yr, marking a return to early Holocene conditions.

Despite its incomplete stratigraphic record of coastal evolution during the middle and early Holocene, it is still the most complete record for the Gulf Coast. Bay evolution, as recorded within the offshore Trinity and Sabine incised valleys, was characterized by periods of bayhead delta and tidal delta expansion, followed by episodes of dramatic landward shifts in these environments. The ancestral Brazos, Colorado, and Rio Grande river deltas and coastal barriers also experienced landward stepping

---

[*]*Present address:* U.S. Army Corps of Engineers, Omaha, Nebraska 68102, USA.

Anderson, J.B., Wallace, D.J., Rodriguez, A.B., Simms, A.R., and Milliken, K.T., 2022, Holocene Evolution of the Western Louisiana–Texas Coast, USA: Response to Sea-Level Rise and Climate Change: Geological Society of America Memoir 221, p. 1–81, https://doi.org/10.1130/2022.1221(01). © 2022 The Authors. Gold Open Access: This chapter is published under the terms of the CC-BY license and is available open access on www.gsapubs.org.

during the early Holocene. The widespread nature of these flooding events and their impact on multiple coastal environments suggests that they were caused by episodes of rapid sea-level rise.

Similar methods were used to study modern bays, including the acquisition of seismic lines and drill cores along the axes of the bays to examine the magnitudes and timing of transgressive events. Results from Lake Calcasieu, Sabine Lake, Galveston Bay, Matagorda Bay, Copano Bay, Corpus Christi Bay, and Baffin Bay reveal that landward shifts in bayhead deltas, on the order of kilometers per century, occurred between 9.8 ka and 9.5 ka, 8.9–8.5 ka, 8.4–8.0 ka, and 7.9–7.5 ka. These results are consistent with those from offshore studies and indicate that punctuated sea-level rise dominated coastal evolution during the early Holocene.

By ca. 7.0 ka, the average rate of sea-level rise in the northern Gulf of Mexico decreased to 1.4 mm/yr, and there was considerable sinuosity of the coastline and variability in the timing of bay and coastal barrier evolution. The diachronous nature of coastal environment migration across the region indicates that sea-level rise played a secondary role to climate-controlled oscillations in river sediment discharge to the coast. At ca. 4.0 ka, the average rate of sea-level rise decreased to 0.5 mm/yr. During this period of slow sea-level rise, coastal bays began to take on their current form, with the exception of changes in the sizes and locations of bayhead deltas caused by changes in sediment supply from rivers. There were also significant changes in the size and configuration of tidal inlets and deltas as a result of barrier growth. The late Holocene was also a time when coastal barriers experienced progradation and transgression on the order of several kilometers. The timing of these changes varied across the region, which is another indication that sea-level rise played a minor role in coastal change during the late Holocene. Instead, barrier evolution during this time was controlled by fluctuations in sand supply to the coast from rivers and offshore sources.

Historical records indicate a dramatic reversal in coastal evolution marked by increased landward shoreline migration of chenier plains and coastal barriers across the region. The main cause of this change is accelerated sea-level rise during this century and diminished sediment supply to the coast. Wetlands are also experiencing rapid change due to their inability to keep pace with sea-level rise, especially in areas where subsidence rates are high. Although direct human influence is a factor in these changes, these impacts are more localized. Coastal change is expected to increase over the next several decades as the rate of sea-level rise increases, the climate in Texas becomes more arid, and more severe storms impact the coast.

INTRODUCTION

We can better prepare for the inevitable response of coastal systems to global climate change by studying how they responded to past changes in climate and sea-level rise. Achieving this objective requires detailed records from a range of coastal environments with sufficient temporal resolution to allow direct correlation among rates of change and sea-level rise and climate variability. Results from research conducted on the western Louisiana and Texas (LaTex) Coast, USA, over the past four decades provides a detailed record of changes in a number of different coastal environments throughout the Holocene, a time when the rate of sea-level rise varied by more than an order of magnitude (Törnqvist et al., 2004a, 2004b; Milliken et al., 2008a; Kolker et al., 2011) and climate was alternating between wet and dry cycles that influenced river discharge and sediment supply to the coast (Livsey et al., 2016; Milliken et al., 2017). Indeed, the current rate of sea-level rise is approaching that of the early Holocene, and west Texas is experiencing drier conditions that are predicted to continue in the near future (Nielsen-Gammon et al., 2020). Coupled with more direct human alteration of fluvial and coastal systems, these changes are causing alarming rates of coastal change.

Located west of the Mississippi Delta, the LaTex Coast (Fig. 1) includes a myriad of coastal environments, including chenier plains, barrier islands and peninsulas, tidal inlets and deltas, fluvial delta headlands, bays, and lagoons (Fig. 2). These low-lying coastal environments are highly vulnerable to rising sea level, changing climate, and severe storm impacts (FitzGerald et al., 2008). The LaTex coastal area differs from the Mississippi Delta,

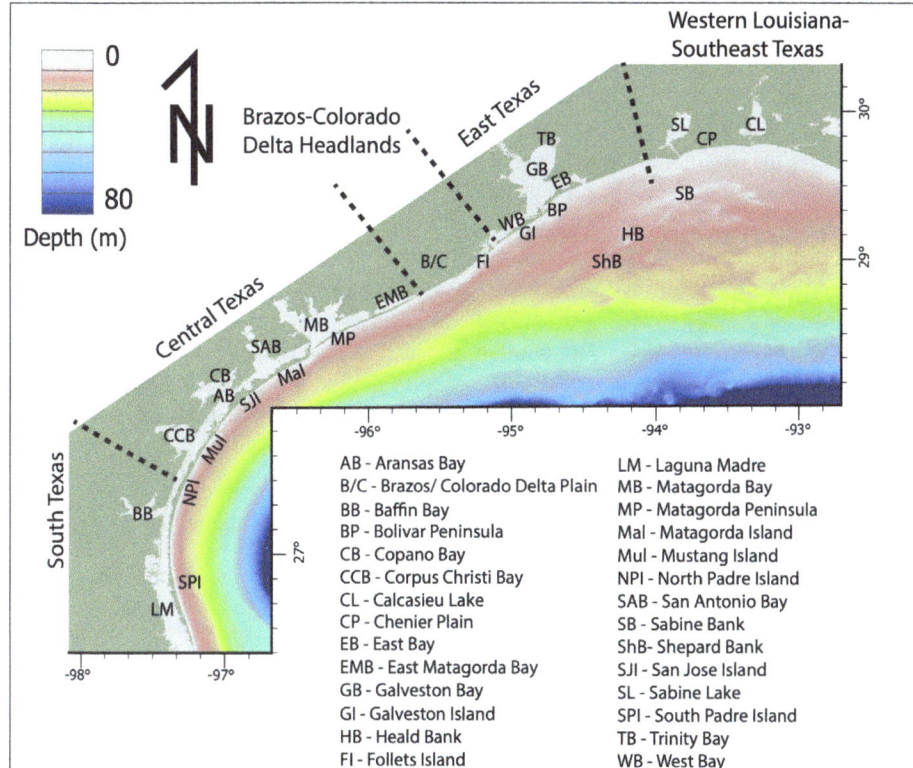

Figure 1. Map of study area shows geographic names and boundaries of coastal segments discussed in text. Also shown is the bathymetric map for the region (base map from GeoMapApp [www.geomapapp.org] / CC BY), which highlights the broad, gentle gradient of the western Louisiana and east Texas continental shelf versus the steeper central and south Texas continental shelf.

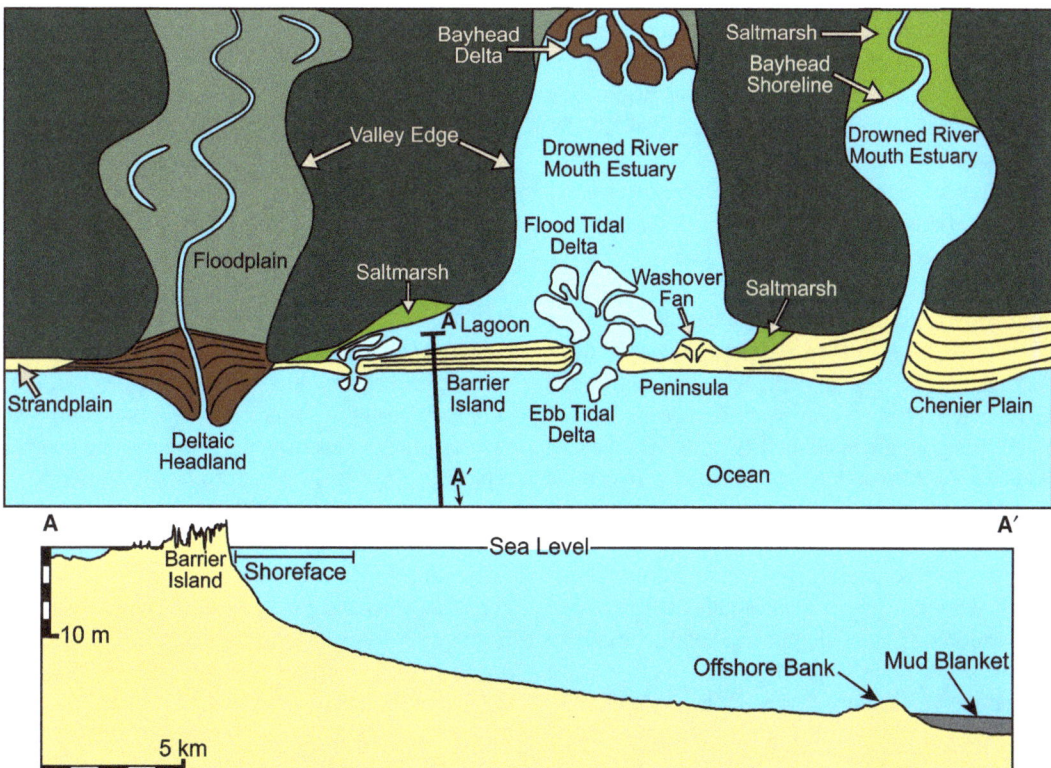

Figure 2. Main coastal environments of the western Louisiana and eastern Texas Coast are shown, and a representative onshore to offshore section below illustrates significant geomorphic and depositional features.

where deltaic processes dominate, and from the Mississippi, Alabama, and northwestern Florida coasts, which are characterized by narrow coastal plains bounded by areas of relatively high relief and networks of dendritic coastal plain and piedmont rivers that have nourished the coast with sand (Greene et al., 2007; Hollis et al., 2019).

For a number of reasons, the LaTex Coast is an ideal natural laboratory for investigating low-gradient coastal response to global climate change. The adjacent continental shelf contains the best, albeit incomplete, early Holocene geological records of coastal change for the U.S. Gulf Coast, a time when the rate of sea-level rise was of a magnitude similar to the current rate (Anderson et al., 2014). Holocene deposits on the continental shelf of the Mississippi, Alabama, and northwestern Florida coasts have been eroded and reworked into a sheet sand that covers much of the continental shelf in the eastern Gulf of Mexico (McBride et al., 1999). In Louisiana, delta lobe shifting during the Holocene has resulted in a complex stratigraphic record of delta evolution that is too thick to be sampled by conventional coring methods (Berryhill, 1987).

Climate varies widely across the region, and this variability, coupled with large differences in river drainage basin area, results in a wide range in the mean annual sediment discharge of rivers (Fig. 3). The largest rivers in the study area are the Brazos, Colorado, and Rio Grande rivers. Their drainage basins span multiple climate zones and are susceptible to large-scale climate variability. Currently, high-discharge events for these rivers are caused by meteorological fronts that originate in the Pacific Ocean. In contrast, smaller rivers in the area are influenced by more regional climate variability with high-discharge events that are often driven by tropical disturbances originating in the Gulf of Mexico, Caribbean, and Atlantic (Zhu et al., 2015). Holocene paleoclimatic records for Texas reveal a highly variable climate that has been dominated by shifts from warm/dry to cool/wet conditions (Humphrey and Ferring, 1994; Nordt et al., 1994, 2002; Toomey et al., 1993; Wong et al., 2015; Livsey et al., 2016). In addition, the physiography of the LaTex coastal area varies. Southwestern Louisiana and southeastern Texas are characterized by a wide, low-gradient Pleistocene coastal plain and broad, low-gradient continental shelf, whereas central and south Texas is characterized by a narrower Pleistocene coastal plain and relatively steep continental shelf (Fig. 1). The central Texas shelf is a relatively deep embayment situated between the ancestral Colorado and Rio Grande deltas. As we will see, these differences have strongly influenced the evolution of coastal environments. For these reasons, the study area is separated into five coastal regions: western Louisiana and southeast Texas, east Texas, Brazos and Colorado fluvial–deltaic headlands, the central Texas Coast, and the south Texas Coast (Fig. 1).

We use marine oxygen isotope stages (MIS) as a chronostratigraphic framework, as shown in Figure 4. Our discussion begins with the pre-Holocene evolution of fluvial-dominated deltas and strand plains, from late Pleistocene MIS 5e (ca. 120 ka) to the beginning of the Holocene (11.7 ka), to gain an appreciation for the full range of coastal responses to climate and sea-level change across a complete glacial-eustatic cycle (Fig. 4A). This perspective is important because deposition and erosion during these times helped shape the physiography of the continental shelf and provide the foundation on which Holocene coastal systems evolved. The Pleistocene deltas have also been an important source of sand for the modern coastal system. Next we focus on coastal evolution during the Holocene, the period for which sea-level change in the western Gulf of Mexico is best constrained (Fig. 4B) and a period for which the stratigraphic record of coastal evolution is most complete. This is followed by a discussion of the factors that influenced coastal change and concludes with a section that addresses current coastal change as the sediment-starved coast experiences a dramatic increase in sea-level rise.

STUDY AREA

The study area spans nearly 1000 km of coastline with a spectrum of modern environments (Fig. 1). It also spans four climate zones: humid in western Louisiana and eastern Texas, wet subhumid in east central Texas, dry subhumid in central Texas, and semiarid in southern Texas. Mean annual temperatures vary slightly across the region, but mean annual precipitation ranges from 50 cm/yr to 175 cm/yr (Fig. 3). The broad range in precipitation is an important influence on vegetation type and density and on the salinity structure of bays. In general, vegetation land cover decreases from east to west, and bay salinities increase from north to south along the coast. The strong precipitation gradient is also an important influence on river water and sediment discharge, although precipitation is overshadowed by drainage basin size and geology, which vary widely across the region (Anderson et al., 2016; Milliken et al., 2017; Fig. 3). Given this extreme climate gradient, it is not surprising that Holocene paleoclimate records show significant spatial and temporal variability over centennial to millennial timescales (Wong et al., 2015).

Tides are predominantly diurnal with an average fair-weather tidal range of less than 0.5 m at the coast. Tidal influence on coastal circulation is minor, except within tidal inlets and bays where it can be amplified. Otherwise, winds dominate coastal circulation. During warmer months, coastal currents are controlled by prevailing winds from the southeast and near-shore waves that typically range from 30 cm to 60 cm in height with wave periods of 2–6 s (Morton, 1994). These southeasterly winds result in currents that flow along the coast and transport sediment with them (Fig. 5). Because of the curvature of the coast, longshore currents flow from east to west along the upper Texas Coast and from south to north along the south Texas Coast. The convergence of these currents along the central Texas Coast has had a significant influence on coastal evolution, with the central Texas Coast receiving greater sand supply from the longshore transport system (Lohse, 1955; Curray, 1960; Morton, 1979).

The LaTex coastal region is influenced by meteorological fronts that approach from the west, mostly during late autumn

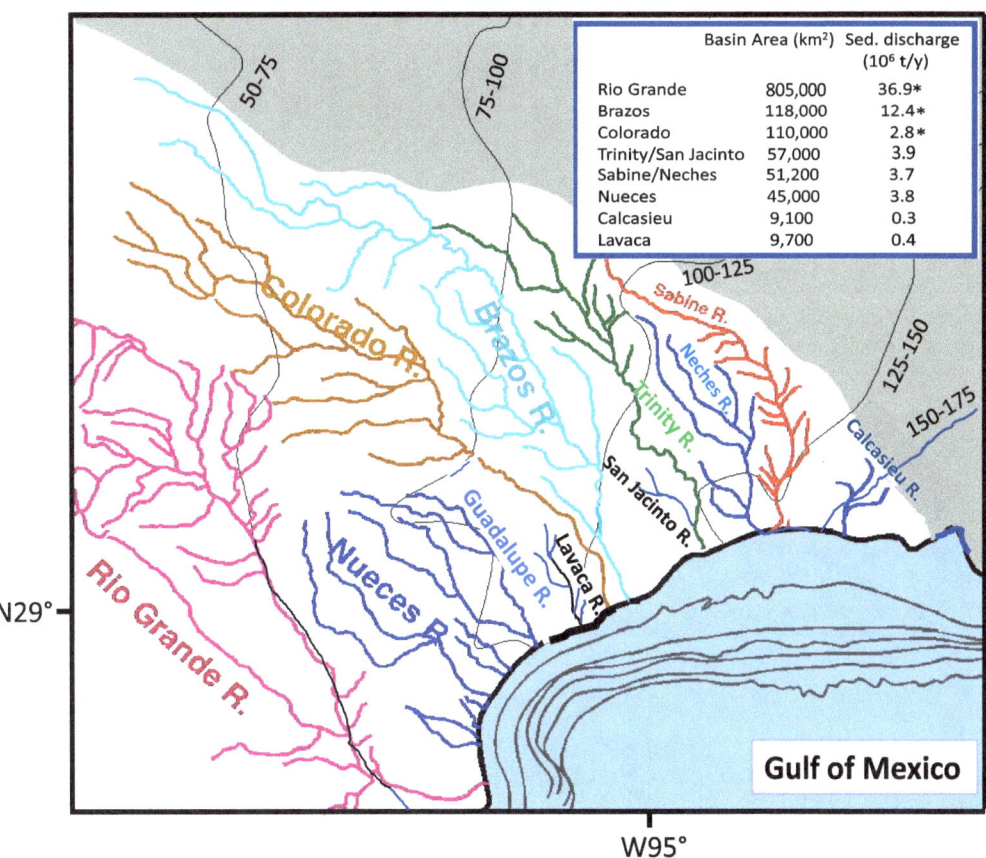

Figure 3. Map shows drainage basins for rivers that flow into the LaTex Coast (modified from Anderson et al., 2004). Thin lines show mean annual precipitation contours in cm/yr. The inset table provides drainage basin areas and long-term sediment discharge as derived using the BQART equation (Syvitski and Milliman, 2007) for those rivers that currently nourish delta headlands and bayhead deltas. Asterisk denotes sediment discharge rates from Anderson et al. (2016). Other rates are from Milliken et al. (2017).

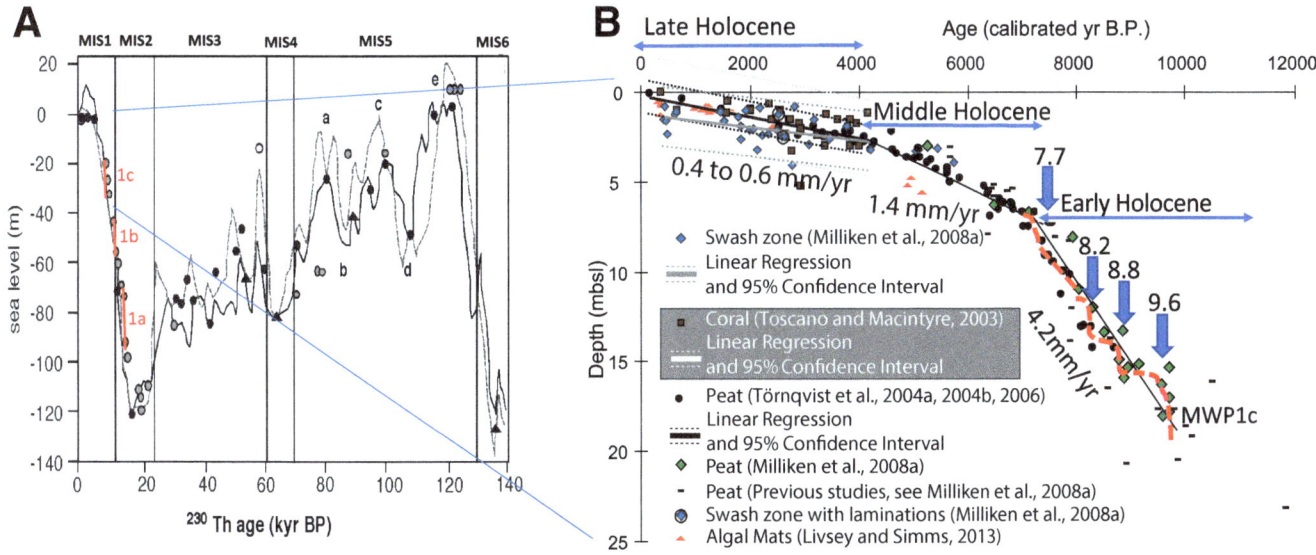

Figure 4. (A) Sea-level curve for the last glacial eustatic cycle with marine isotope stages labeled MIS 6 through MIS 1 (data from Labeyrie et al., 1987; Shackleton, 1987; Bard et al., 1990; Chappell et al., 1996; Anderson et al., 2004). (B) Holocene sea-level curve for the northwestern Gulf of Mexico with various data sources provided (modified from Anderson et al., 2014). For error bars, see referenced studies. Note, Livsey and Simms (2013) is not included in the linear regression analyses in panel B. Blue arrows designate punctuated sea-level events.

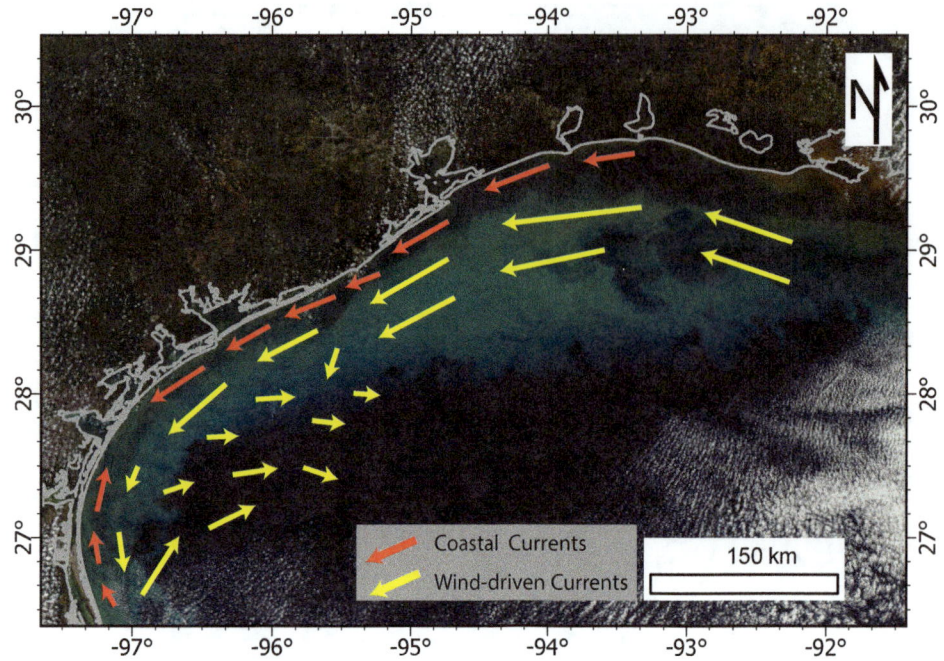

Figure 5. Dominant circulation patterns in the northwestern Gulf of Mexico are shown. Red arrows depict prevailing longshore current directions, and yellow arrows depict highly generalized wind-driven surface currents.

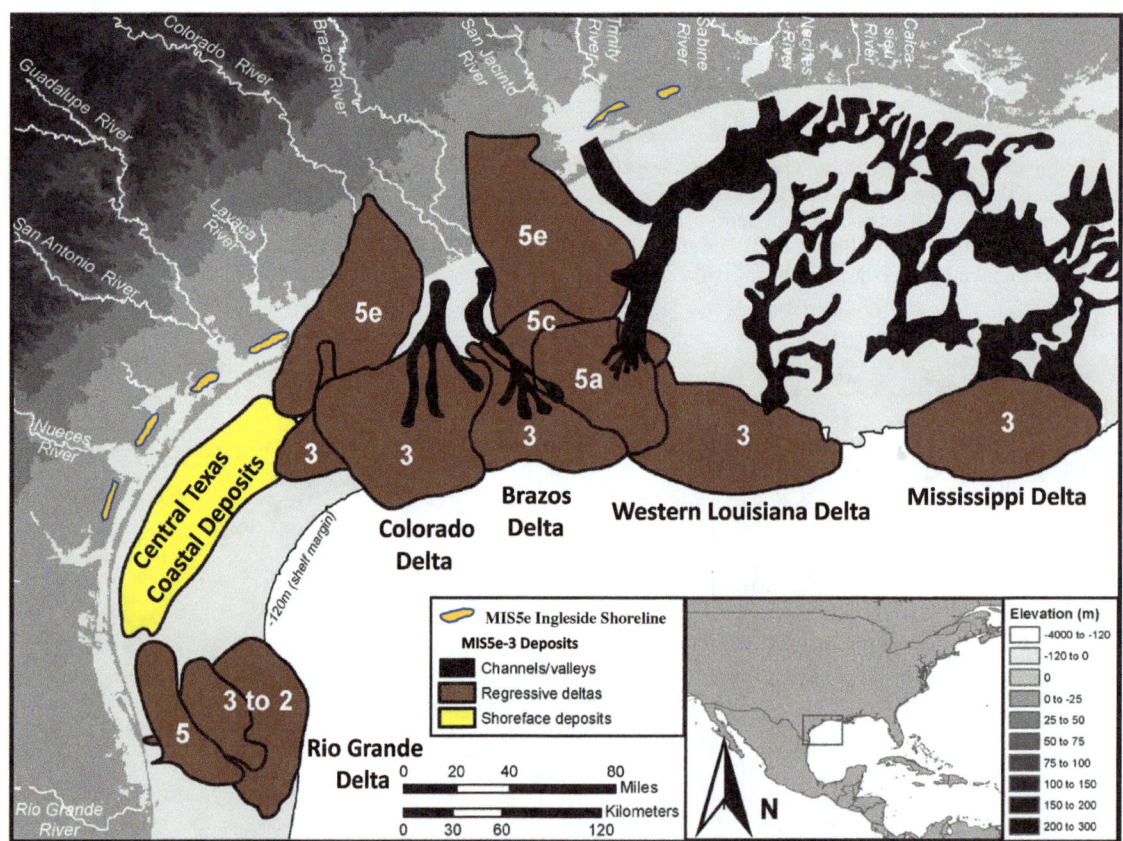

Figure 6. Paleogeographic map of the LaTex continental shelf shows major deltas in brown, fluvial channels and incised valleys in black, and central Texas coastal deposits in yellow. Channels on the western Louisiana continental shelf are from Suter and Berryhill (1985). The numbers on the different deltas indicate the marine isotope stages (Fig. 4A) of the deltas as determined by Abdulah et al. (2004), Wellner et al. (2004), and Banfield and Anderson (2004). Figure was modified from Anderson et al. (2016). Also shown are approximate locations of onshore exposures of the MIS 5e Ingleside Paleoshoreline.

and early spring. Approaching fronts are marked by winds from the west, which shift to the north and east as they move eastward along the coast. As the front passes through the region, longshore currents initially flow toward the east but diminish as winds shift to a more offshore direction and then shift back to their predominant southeastern flow direction with the trailing edge of the front. These fronts also influence estuarine circulation and sediment erosion and transport, which is more complicated due to the highly variable shape and orientation of the bays. Continental shelf circulation is dominated by winds, with tides being a secondary influence. Wind-driven circulation on the continental shelf is mainly from east to west, which results in the transport of fine-grained sediments from the Mississippi River toward the west and along the Texas Coast (Fig. 5). A significant portion of this sediment is deposited on the central Texas continental shelf in the form of the Texas Mud Blanket (Weight et al., 2011).

Over the past few thousand years, intense hurricanes have impacted the south Texas Coast with an average landfall probability of 0.46% (Wallace and Anderson, 2010). Storm magnitudes appear to be increasing in response to increasing Gulf of Mexico sea-surface water temperatures, although storm tracks are regulated by large-scale meteorological processes, such as El Niño–Southern Oscillation and the North Atlantic Oscillation (Wallace et al., 2014). It remains unclear how hurricane activity will respond to anthropogenic influences, although there is some consensus that hurricane intensity will increase but the overall frequency of storms could decrease (Knutson et al., 2010; Villarini and Vecchi, 2013). In recent years, several storms have intensified rapidly before reaching the coast, and this is attributed to warmer sea-surface temperatures (Russell et al., 2020).

The physiography of the Texas continental shelf reflects differences in subsidence, sediment supply, and sediment dispersal that have occurred mainly since the early Miocene in response to an eastward shift in main sediment source and depocenter (Galloway et al., 2011). The Rio Grande was a larger source of sediment to the basin prior to the early Miocene; since that time, much of the sediment derived from denudation of the southern Rocky Mountains has been deposited within the Rio Grande Trough and the Mississippi has become the dominant sediment supplier to the basin. Since the early Miocene, the western Louisiana–east Texas continental shelf has prograded ~200 km while the central and south Texas shelf has prograded less than half that distance. During the Pleistocene, large deltas formed by the Mississippi, Brazos, Colorado, and Rio Grande rivers advanced and retreated across the continental shelf in response to ca. 100 ka glacial-eustatic cycles, which resulted in the current physiography of the continental shelf (Anderson et al., 2004, 2016). Shoreface to inner shelf profiles vary along the coast, with the western Louisiana and east Texas coasts having gentle profiles relative to the central and south Texas Coast. This reflects spatial differences in substrate lithology, sediment supply, and wave climate.

Bays within the study area generally fall into two categories: shallow, shore-parallel bays and lagoons on the landward side of coastal barriers and deeper bays oriented more perpendicular to the coast that occupy incised river valleys (Fig. 2). The former have relatively shallow Pleistocene surfaces with relatively thin overlying Holocene sediments while the latter have thick Holocene sediment fills that record variations in sediment supply and sea-level change, as well as changes in the coastal barriers and tidal inlets that isolate these bays from the open gulf (Anderson and Rodriguez, 2008).

Sea-Level History

During the previous (MIS 5e) highstand, sea level in the northern Gulf of Mexico was as much as 7.2 m higher than at present, and the shoreline was located between 10 km and 20 km landward of its current location (Fig. 6; Simms, 2021). Following this highstand, sea level fell episodically, with fluctuations of a few tens of meters over time periods of roughly 20,000 years (Fig. 4A). The maximum lowstand in the western Gulf of Mexico was about −90 m to −100 m and occurred at the same time as the maximum global lowstand between 22 ka and 19 ka (Fig. 4A; Yokoyama et al., 2000; Simms et al., 2007a; Clark et al., 2009). During the subsequent ~10 k.y., sea level rose nearly 70 m (Fig. 4A) and at a rate that was typically greater than 10 mm/yr and at times was greater than 20 mm/yr, driven mainly by deglaciation in the Northern Hemisphere (e.g., Curray, 1960; Fairbanks, 1989; Bard et al., 1990, 1996; Peltier and Fairbanks, 2006; Simms et al., 2007a). The handful of available sea-level index points from the Gulf of Mexico for this period appear to generally follow the global eustatic curve, albeit at slightly higher elevations in western Louisiana that are assumed to reflect Glacial Isostatic Adjustment (GIA) influence caused by the retreat of the North American Ice Sheet (Simms et al., 2007a; Kuchar et al., 2018). Coralgal reef terraces offshore central Texas record punctuated sea-level events of around 5 m to 6 m magnitude over decadal to century timescales between ca. 15.5 ka and ca. 11.5 ka (Khanna et al., 2017) as the shoreline migrated across the outer continental shelf. Otherwise, late Pleistocene sea-level records for the western Gulf are poorly constrained, mainly because paleoshorelines for this time have been either eroded or buried beneath Holocene marine mud.

The Holocene sea-level history of the northwestern Gulf of Mexico has been the subject of several recent investigations (Törnqvist et al., 2004a, 2004b, 2006; Milliken et al., 2008a; Livsey and Simms, 2013) and was used to construct the composite sea-level curve shown in Figure 4B. Different methods were used to acquire sea-level datums and correct for subsidence, as described by the authors. This curve is generally accurate to within +/− 2 m for the early half of the Holocene, which allows reasonable age constraints for paleoshorelines and deltas for which radiocarbon ages are lacking. The curve shown in Figure 4B was analyzed using three linear regressions and indicates three steps in the rate of rise at ca. 10 ka to ca. 7 ka (4.2 mm/yr), ca. 7 ka to ca. 4 ka (1.4 mm/yr), and ca. 4 ka to Present (0.5 mm/yr). The rate for the past 2 ka is well constrained (~0.3 mm/yr) using data from Baffin Bay in south Texas (Livsey and Simms,

2013). Since we rely somewhat on the sea-level curve to constrain the timing of coastal evolution, we apply unofficial designations of early, middle, and late Holocene to these time intervals for ease of discussion (Fig. 4B).

Historical sea-level records from tide gauges are supported by satellite data and indicate that the rate of eustatic rise significantly increased over the past two centuries, from a global average of 1.4 mm/yr to the current rate of around 3.6 mm/yr and potentially as high as 4.0 mm/yr (Chen et al., 2017; Nerem et al., 2018; Dangendorf et al., 2019; Wang et al., 2021; Oppenheimer et al., 2019). In the northern Gulf of Mexico, the current average rate of rise after correcting for vertical land motion, steric affects, and GIA exceeds 3.0 mm/yr (Wang et al., 2021), which is a multi-fold increase over the late Holocene rate of ~0.5 mm/yr. This historical acceleration of sea-level rise is consistent with results from the U.S. East Coast that show a significant increase in the rate of rise since the nineteenth century (Englehardt et al., 2011; Kemp et al., 2011), although the actual rate there is different from the rate at the western Gulf of Mexico due mainly to GIA effects.

The historical acceleration of sea-level rise is projected to continue, with estimates ranging from 2.6 mm/yr to 8.1 mm/yr by the end of this century (Bamber et al., 2019; Oppenheimer et al., 2019). The greatest uncertainty in these predictions has to do with future contributions from ice sheets. Another uncertainty with regard to sea-level rise in the LaTex region relates to land subsidence.

In general, long-term subsidence decreases from east to west across the Gulf of Mexico Basin, reaching a maximum offshore Louisiana and minimum on the Florida Platform. Along the western Gulf Coast, relative sea-level rise currently varies by an order of magnitude due to variable subsidence across the region. Törnqvist et al. (2008) used more than 100 shallow boreholes to determine a linear relationship between the compaction rate of organic-rich sediment and overburden thickness. Subsequently, Jankowski et al. (2017) used 258 rod surface elevation table-marker horizon stations to measure relative sea-level rise in Louisiana. Their results showed that 58% of the western Louisiana Chenier Plain is being inundated by relative sea-level rise, and at least 60% of the total subsidence occurs within the upper 5–10 m of the sediment column.

In Texas, Pleistocene sediments are locally exposed onshore and in intertidal areas and occur at shallow depths in shore-parallel bays, which indicates that long-term, regional-scale rates of subsidence are minimal (< 0.1 mm/yr) (Paine, 1993; Simms et al., 2013). Short-term (decadal to century-scale) subsidence generally exceeds the long-term rate of tectonic subsidence by at least an order of magnitude, with results from tide-gauge records indicating rates in the range of 0.8–3.3 mm/yr, assuming a eustatic rate of 3.0 mm/yr (Fig. 7). The higher rates occur in bays that occupy incised valleys, which indicates that subsidence is mainly due to compaction of thick (up to 60 m)

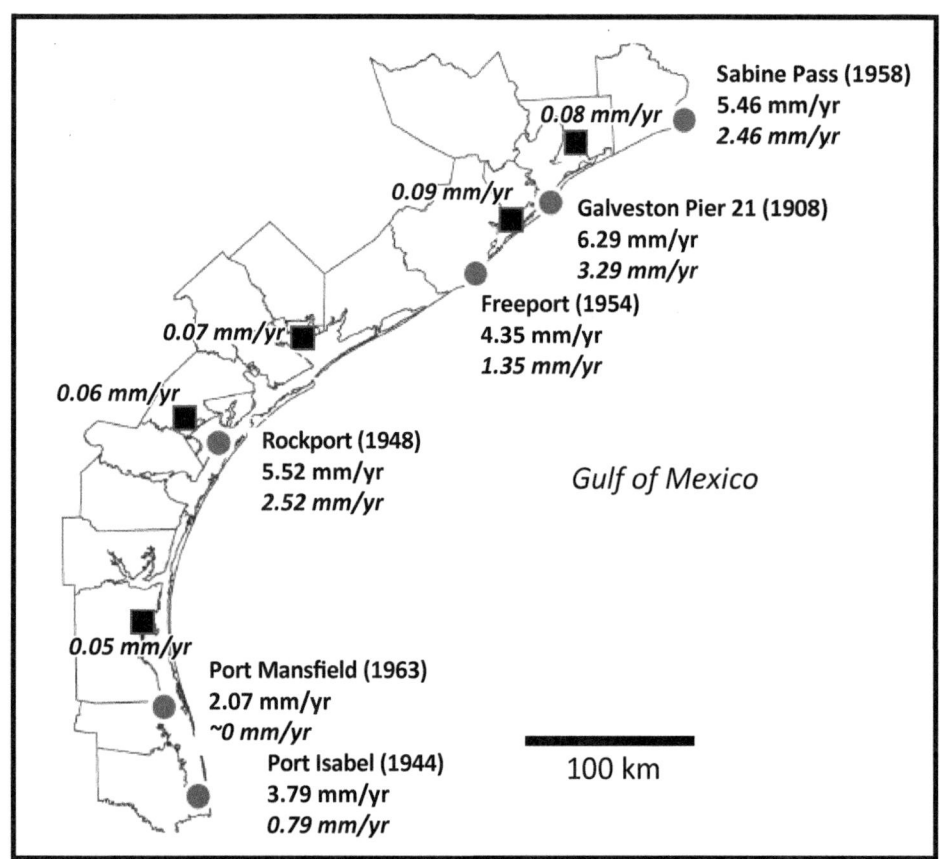

Figure 7. Relative sea-level rise based on tide gauge records for Texas (gray circles) is shown. Numbers in italics are estimated subsidence rates derived by subtracting a eustatic rate for the Gulf of Mexico (3.0 mm/yr) from the tide gauge measurement rate (data from the National Oceanic and Atmospheric Administration). The variability in subsidence across the region is attributed mainly to differences in the thickness of Holocene sediments (i.e., compaction) and anthropogenic processes (i.e., fluid withdrawal). Also shown are core locations (black squares) where Pleistocene sediments were sampled at shallow subsurface depths in sediment cores or where they are exposed onshore and thus indicate low rates of subsidence.

Holocene sediments within these valleys (Anderson et al., 2014). High rates of localized subsidence can also result from subsurface fluid extraction and surface faulting. Historical localized subsidence rates due to combined sediment compaction and fluid withdrawal range between 3.0 mm/yr and 22.0 mm/yr (Paine, 1993; Meckel et al., 2006; Morton et al., 2006; Kolker et al., 2011). Literally hundreds of active growth faults occur along the LaTex Coast and are known to result in displacement of the ground surface in excess of 1 m per century (e.g., Verbeek, 1979; McCulloh and Heinrich, 2012). Since this displacement is typically downward in an offshore direction, it results in abnormally high rates of subsidence.

Studies in Louisiana have shown that neotectonics have had an important, albeit spatially restricted, influence on coastal subsidence (Chan and Zoback, 2007; Yuill et al., 2009; Shen et al., 2017). Relatively few detailed studies have focused on the influence of growth faults on coastal environments in Texas, although aerial images show linear features all along the Texas Coast that are likely growth faults. One of the most detailed studies of one of these faults was conducted by Feagin et al. (2013), who measured a maximum offset rate of ~0.75 m over the past ~40–50 years near the mouth of the Colorado River.

DATA SETS AND METHODS

Because this research spanned nearly four decades, our equipment and analytical methods evolved over time. In all, we acquired thousands of kilometers of high-resolution seismic data and hundreds of sediment cores and radiocarbon dates spanning the LaTex Coast and continental shelf. This research was conducted using three research vessels, the *R/V Matagorda*, the *R/V Lone Star*, and the *R/V Trinity*, along with numerous small vessels used in inland waters. The *R/V Matagorda* is a 13 m shallow draft vessel used to acquire high-resolution (boomer) seismic data from area bays. The *R/V Lone Star* is a 22 m vessel used mainly for offshore seismic surveys and coring using a pneumatic hammer that could retrieve cores up to 5 m long in water depths of up to 12 m. The *Lone Star* was equipped with an air compressor, enabling the use of air and water guns to complement high-resolution seismic data collected with boomer and Compressed High Intensity Radar Pulse (CHIRP) systems. The *R/V Trinity* is a 7 m drill barge equipped with a water well-type of drilling rig capable of core sampling to sub-bottom depths of 30 m with an average recovery rate of 80%. It was used exclusively for work in bays. The poorest recovery was in the upper few meters of the sediment column and in the lower sand-dominated fluvial/deltaic section. Surface sediment recovery was supplemented with shallower coring techniques. Onshore studies relied heavily on vibra-coring, which proved capable of sampling to sub-bottom depths of 8 m. Sediment cores were cut and described soon after acquisition and placed in refrigerated storage until more detailed analyses were performed. Routine sedimentological analyses included the identification of sedimentary structures, grain size, and macro- and micro-faunal analyses.

An extensive grid of seismic data was acquired on the continental shelf and within bays using different methods to maximize stratigraphic resolution at different water depths, stratigraphic thicknesses, and sediment types (Fig. 8). Data from coastal bays were acquired using an EG&G Uniboom system and Datasonics CAP 6000 and Edgetech SB-216 CHIRP systems. Onboard processing included band-pass filtering and automatic gain control. Offshore seismic data were acquired using a variety of seismic sources, including a 50 cu in Generator-Injector (GI) air gun, 15 cu in water gun, 1 kJ sparker array, and a Uniboom system. All are single-channel data, and most were digitally acquired and processed using band-pass filters and gain adjustment.

We discuss two different types of shoreline change analysis. Paine et al. (2012, 2021) present net rates of shoreline change for the Texas Coast. We also rely on current natural average migration rates (Anderson et al., 2014), where clear anthropogenic impacts for each barrier island, mostly near inlets, have been removed.

Radiocarbon Dating

During the 1980s and early 1990s, we used commercial labs for radiocarbon analyses. The vast majority of our radiocarbon ages were acquired after the mid-1990s and were measured at the University of California Irvine Keck Carbon Cycle Accelerator Mass Spectrometer (KCCAMS) and National Ocean Sciences Accelerator Mass Spectrometry (NOSAMS) radiocarbon facility. All radiocarbon ages are reported as calibrated yr B.P. (cal yr B.P.), where present is 1950 CE, and approximate ages are reported in kilo-years (ka). The greatest uncertainty in our radiocarbon ages of estuarine shells comes from uncertainties in the carbon reservoir, which is known to range across the study area from ~760 years in central Texas (Aten, 1983; Kibler, 1999; Simms et al., 2007b, 2008; Milliken et al., 2008a; Rice et al., 2020) to ~200 years in Baffin Bay, south Texas (Simms et al., 2009). Using a local reservoir where available, or the standard 400 yr marine reservoir correction where unavailable, the ^{14}C ages were converted to calibrated calendar years at the 95.5% confidence interval (2 sigma ranges) with Marine13 using the latest version of the CALIB.Rev. 5.0 program (Reimer et al., 2013). Older publications used calibration curves available during the original time of publication; however, any differences are relatively minor. Ages from Follets Island were acquired using the continuous flow gas bench accelerator mass spectrometer method at the Woods Hole Oceanographic Institution's NOSAMS facility, which results in a slightly higher range of uncertainty. All ages greater than 30 ka, regardless of methodology, are considered radiocarbon dead.

RESULTS

Late Pleistocene Coastal Evolution

The Holocene evolution of the LaTex Coast was significantly influenced by events that occurred during the MIS 5e through MIS 1 sea-level cycle. During this time, large offshore

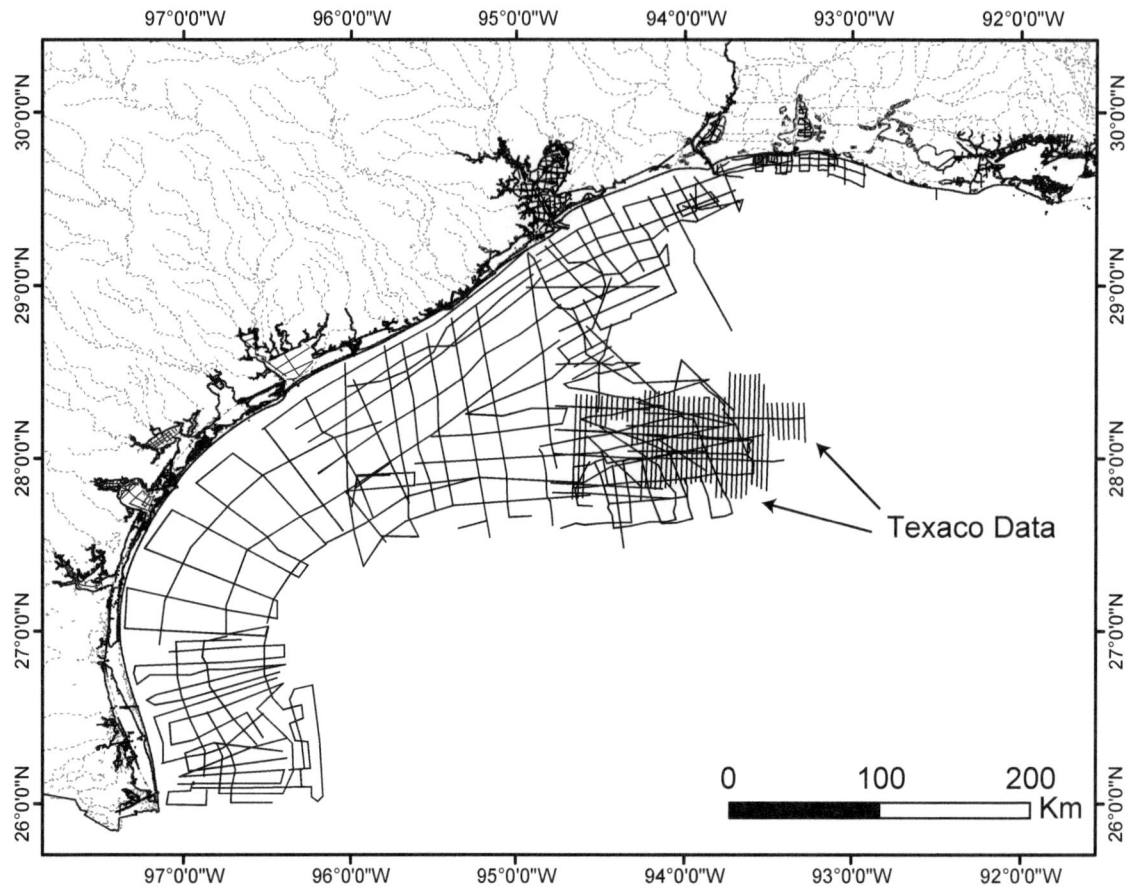

Figure 8. Map shows high-resolution seismic data used for offshore mapping of deltas, incised river valleys, paleoshoreline features, and the Texas Mud Blanket using a variety of methods. These data were augmented by sparker data from previous studies by the U.S. Geological Survey (Berryhill, 1987) on the western Louisiana shelf (not shown) and Texaco Oil Company to study the major depositional features on the continental shelf. Also shown are seismic tracklines used to study onshore bays. Figure was modified from Anderson et al. (2016).

deltas formed. These deltas would ultimately provide much of the sand that nourished cheniers, coastal barriers, and peninsulas during the Holocene. It was also the period of formation of the relief on which Holocene coastal environments evolved.

MIS 5–2 Sea-Level Fall and Lowstand

During the MIS 5e highstand, the Texas shoreline, which is known as the Ingleside paleoshoreline, was situated ~30 km landward of the modern shoreline along the east Texas Coast but less than half that distance from the modern central Texas shoreline (Otvos and Howat, 1996; Otvos, 2001, 2005a, 2005b, 2018; Simms et al., 2013; Simms, 2021; Fig. 6). This reflects the different coastal plain gradients of these areas. The Ingleside paleoshoreline is a prominent topographic feature that averages 6–8 m elevation and, along much of the Texas Coast, is a prominent step marking the northern boundary of the modern coastal plain.

During the MIS 5–3 fall in sea level, the Mississippi, Red, Calcasieu, Sabine, and Neches rivers constructed a vast channel network on the western Louisiana continental shelf (Fig. 6). During MIS 3, the Mississippi River constructed two large, fluvial-dominated deltas, one on the western Louisiana shelf (Suter and Berryhill, 1985) and another, the western Louisiana Delta, on the east Texas continental shelf (Anderson et al., 1996; Wellner et al., 2004). The Brazos, Colorado, and Rio Grande rivers also nourished large, fluvial-dominated deltas that prograded across the continental shelf in step with the MIS 5–3 sea-level fall (Fig. 6). Each phase of delta growth was interrupted by a transgressive phase during relatively brief episodes of sea-level rise. These transgressive events were generally less than 10,000 years in duration, and the associated sea-level rises were on the order of a few tens of meters (MIS 5e–d, MIS 5d–c, MIS b–a, and MIS 4–3; Fig. 4A) (Abdulah et al., 2004; Banfield and Anderson, 2004; Anderson et al., 2016). With each new phase of progradation, sediment recycled from older, more landward deltas contributed to the sediment supply for later episodes of delta development (Anderson et al., 2016). The end result was widespread and thick MIS 3–2 shelf-margin deltas and associated slope fan systems.

No large rivers discharge into the central Texas Coast, which explains the absence of MIS 5–3 deltas on the central Texas shelf.

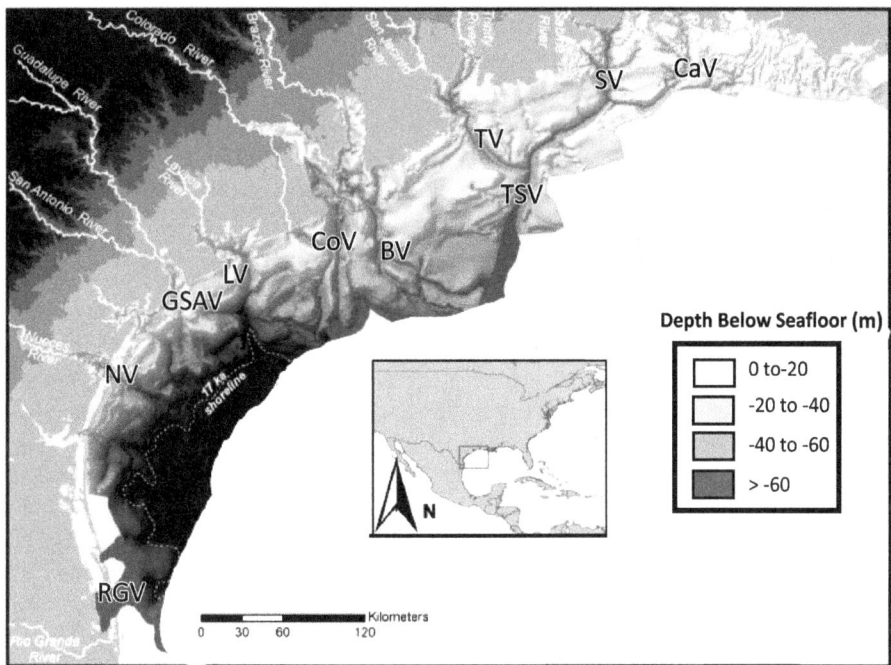

Figure 9. Digital elevation map shows the marine oxygen isotope stage (MIS) 2 Pleistocene surface. CaV—Calcasieu valley, SV—Sabine valley, TV—Trinity valley, TSV—Trinity-Sabine valley, BV—Brazos valley, CoV—Colorado valley, LV—Lavaca valley, GSAV—Guadalupe–San Antonio valley, NV—Nueces valley, and RGV—Rio Grande valley. Modified from Anderson et al. (2016).

Instead, an extensive strand plain, bounded to the north by the Colorado Delta and to the south by the Rio Grande Delta, prograded across the inner shelf as sea-level fell (Eckles et al., 2004; Fig. 6). Sands nourishing this strand plain were mainly derived from the erosion of flanking deltas. Unlike the deltaic areas to the north and south, coastal deposits filled only a portion of the accommodation formed by subsidence, and thus they contributed to the embayment in the central Texas shelf (Fig. 1).

During the MIS 2 sea-level lowstand, rivers and streams eroded the subaerially exposed continental shelf to create the irregular Pleistocene surface on which Holocene coastal systems were established (Simms et al., 2006, 2007a; Anderson et al., 2014; Fig. 9). Differences in fluvial geomorphology between the western Louisiana and east Texas continental shelf are largely a result of the greater size and discharge of Louisiana rivers and flatter shelf gradients offshore Louisiana, which resulted in more frequent aggradation and avulsion than at east Texas rivers and higher rates of shelf mud accumulation (Suter and Berryhill, 1985; Coleman and Roberts, 1988a, 1988b; Anderson et al., 1996). The lowstand physiography of the western Louisiana continental shelf is not well constrained, but it is known that by MIS 2 time the MIS 3 western Louisiana Delta had been abandoned and the Mississippi River shifted to the east, where it cut a deep valley connected to the Mississippi Canyon. To the west, the Calcasieu and Sabine rivers merged on the inner shelf and flowed roughly parallel to the coast for ~100 km before merging with the Trinity/Sabine River (Fig. 9). To the west of the Trinity and Sabine incised valleys, the MIS 2 subaerial exposure surface includes prominent valleys cut by the Brazos and Colorado rivers.

Berryhill (1987) first mapped the lowstand fluvial valleys on the central and south Texas continental shelf using a grid of high-resolution seismic (sparker) profiles. This work showed that the Lavaca, San Antonio, and Mission incised valleys merge on the middle shelf to form one large valley that extends southward and parallel to the coast and merges with the Nueces valley (Fig. 9). Later work by Eckles et al. (2004) showed that this coast-parallel valley follows the seaward edge of the MIS 3 central Texas strand plain, which extends seaward to −60 m. Simms et al. (2007b) argued that the abrupt change in shelf gradient at −60 m, from a low-gradient inner shelf to a ramp-like middle and outer shelf, resulted in minimum fluvial incision of the outer shelf during the late stages of sea-level fall and that the outer portions of these valleys were eroded during transgression.

Banfield and Anderson's (2004) time structure map of the south Texas MIS 2 subaerial exposure surface shows one large incised valley associated with the Rio Grande. This valley increases in width and depth in a seaward direction (Fig. 9) and is filled with more than 140 m of MIS 2–1 sediments at the shelf margin, which indicates that the Rio Grande remained a significant source of sediment to the Gulf of Mexico Basin during the MIS 2 lowstand and subsequent transgression.

MIS 2–1 Transgression

During the rapid sea-level rise between ca. 18 ka and ca. 10 ka, the rate averaged +4.0 mm/yr to +8.0 mm/yr and at times was between +10 mm/yr and +20 mm/yr (Fig. 4A). During this time interval, the shoreline in the western Gulf of Mexico migrated rapidly across the outer continental shelf, eroding coastal deposits and burying them beneath marine mud. Thus, only a sparse stratigraphic record of coastal evolution exists on the continental shelf. The exceptions are bathymetric banks interpreted as remnants of former coastal barriers, estuarine deposits within the Trinity/Sabine/Calcasieu incised river valleys, and the transgressive Brazos, Colorado, and Rio Grande deltas. Here, we discuss results from studies of offshore sand banks and the Trinity/Sabine/Calcasieu incised valleys. The offshore Holocene deltas are discussed later in this Memoir.

Sand Banks

Three prominent bathymetric highs (Sabine, Heald, and Shepard banks) exist on the inner continental shelf within the eastern portion of the study area that have previously been interpreted as former shoreline deposits (Curray, 1960; Nelson and Bray, 1970; Fig. 1). Our investigation of these banks relied on high-resolution seismic data and cross-bank transects of vibracores to examine their evolution (Rodriguez et al., 1999, 2004). All three banks have similar stratigraphic, sedimentological, and paleontological characteristics, which indicates similar development. Of the three, Sabine and Heald banks were most thoroughly studied and are the only banks for which radiocarbon ages exist. We begin our discussion with Sabine Bank, which formed during the mid–late Holocene, because it is the most thoroughly studied and thus provides background for interpreting the older Heald and Shepard banks.

Rodriguez et al.'s (1999, 2004) study of Sabine Bank included the acquisition of multiple high-resolution seismic and side-scan sonar profiles, 27 sediment cores, and 13 radiocarbon ages (Figs. 10A, 10B, 10C). Seismic profiles across the bank show landward- and seaward-dipping strata (Fig. 10D), which is typical of drowned barriers (Mellett and Plater, 2018). Sediment cores from the landward-dipping strata sampled interbedded mud with sand containing estuarine mollusk shells. These are interpreted as estuarine washover deposits and indicate that the ancestral Sabine barrier was located seaward of the bank, where there is a step in the Pleistocene surface. These estuarine deposits are upwards of 4 m thick, which indicates that the back-barrier bay that occupied the Sabine/Calcasieu paleo-valley was at least this deep at the time Sabine Bank was an active barrier. Thus, this bay provided ample accommodation for washover, facilitating landward migration of the barrier and preservation of the landward-dipping strata. An erosional surface separates these back-barrier deposits from overlying seaward-dipping strata (Fig. 10D). Sediment cores that sampled these strata contained interbedded sand and mud with shoreface to inner shelf mollusk shells, indicating that these strata are shoreface deposits that record an episode of progradation. These progradational deposits are truncated by another erosion surface that is overlain by sand with abundant marine mollusk shells indicative of inner shelf water depths. Side-scan sonar profiles and diver observations revealed that the surface of the bank is covered by bedforms that indicate active reworking of the bank at water depths of −8 m to −10 m (Rodriguez et al., 2004).

Core SB96-1 was collected near the center of the Sabine incised valley, landward of Sabine Bank, and sampled estuarine mud overlain by marine mud (Fig. 10C). Samples from estuarine sediments penetrated by this core yielded ages, in stratigraphic order, of 7440 cal yr B.P., 7265 cal yr B.P., and 7090 cal yr B.P. Thus, a bay occupied this part of the valley during this time.

Samples from back-barrier washover and estuarine deposits beneath Sabine Bank contain shells of *Ostrea equestris*, which lives only in brackish water and implies a restricted bay setting (Parker, 1960). Shells from this unit yielded radiocarbon ages of 4695 cal yr B.P., 3130 cal yr B.P., and 2725 cal yr B.P. (Fig. 10C). Ages from the progradational unit range from 2400–1000 cal yr B.P. These ages came from shells of species *Abra aequalis*, *Mulinia lateralis*, and *Ensis minor*, whose depth ranges span the shoreface to inner shelf. Shells and shell fragments of *Donax*, a species that is abundant in the modern foreshore and upper shoreface, were also present. These combined results indicate that Sabine Bank was an active coastal barrier at ca. 4.7 ka and was experiencing transgression and significant washover (Seismic Facies C, Fig. 10) between ca. 3.3 ka and ca. 2.7 ka, followed by an episode of progradation that may have continued after the coastal barrier was drowned (Seismic Facies B). Six radiocarbon ages from the upper sandy unit of the bank range from 2400–1000 cal yr B.P. By ca. 2.5 ka, the shoreline of Bolivar Peninsula was near its present location, and both the Calcasieu and Sabine Chenier plains had started to develop. This indicates that the Sabine/Calcasieu estuary existed landward of Sabine Bank and that the bank was drowned and left isolated on the continental shelf (overstepped) between ca. 4.7 ka and ca. 2.8. During this transgressive event, the shoreline shifted landward ~30 km within ~2000 years (15 m/yr) at a time when the average rate of sea-level rise was only 0.5 mm/yr (Fig. 4B). This rapid transgression appears to have occurred after the Sabine/Calcasieu estuary was filled with sediments and was facilitated by the low gradient of the coastal plain that separated Sabine Bank from the modern coast (Fig. 11).

Seismic profiles across Heald Bank and Shepard Bank reveal seismic units similar to those of Sabine Bank, including an older landward-dipping seismic unit and a younger seaward-dipping unit, which indicate that a period of transgression and washover was followed by a period of progradation and finally by drowning (Rodriguez et al., 1999). Sediment cores from Heald Bank also sampled lithofacies within a stratigraphic succession identical to that of Sabine Bank.

Eight radiocarbon ages were used to reconstruct the history of Heald Bank. Core HB 93-7 was collected north of the bank and sampled sand and muddy sand with mollusk shells indicative of shoreface and/or ebb tidal delta environments resting on estuarine mud. Radiocarbon ages from the estuarine unit yielded ages of 9525 cal yr B.P., 8355 cal yr B.P., and 7770 cal yr B.P. in stratigraphic order. Cores collected from the southern flank of the bank yielded ages of 8415 cal yr B.P. and 8015 cal yr B.P. in stratigraphic order. These combined ages indicate that an active barrier existed slightly seaward of these locations during the span of these radiocarbon ages.

No radiocarbon ages were acquired from Shepard Bank. The base of Shepard Bank occurs at −18 m to −20 m water depth, which indicates that it was a subaerial barrier between ca. 10.0 ka and ca. 9.0 ka. This implies that the landward shift in the shoreline from Shepard Bank to Heald Bank likely occurred during the 9.6 ka punctuated sea-level rise (Fig. 4B). Thomas Bank is surrounded by water depths of −25 m to −28 m, which indicate that it is a pre-Holocene feature.

Offshore Trinity and Sabine Incised River Valleys

The isolated occurrence of the Shepard, Heald, and Sabine paleoshorelines implies that these features were drowned in place by punctuated sea-level rise. However, based on results from numerical modeling experiments, Ciarletta et al. (2019) argue that even for constant rates of sea-level rise and shelf gradient, stepped retreat of barrier islands can occur. Sea-level curves for the late Pleistocene do indicate episodes of rapid rise, which are dubbed meltwater pulses (Fig. 4A). This is supported by the presence of drowned coralgal reef terraces on the central Texas shelf that, although undated, are at water depths consistent with some of these global meltwater pulses (Khanna et al., 2017). Some of

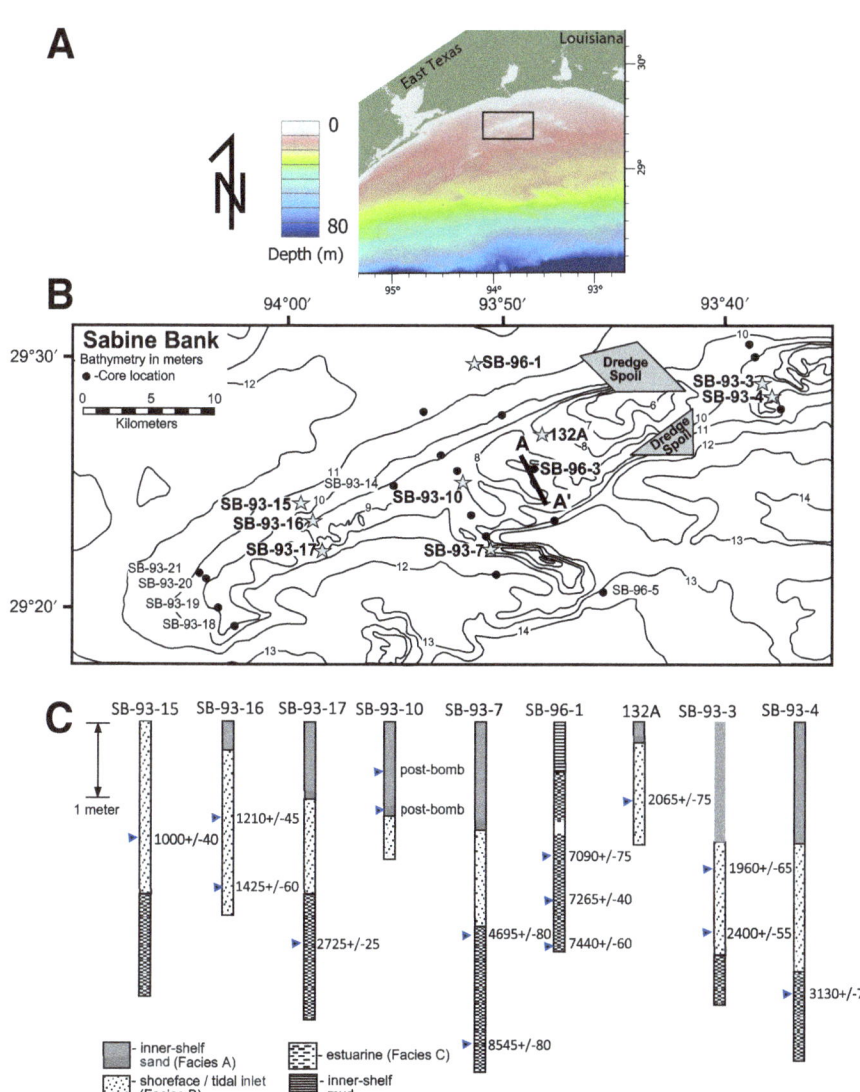

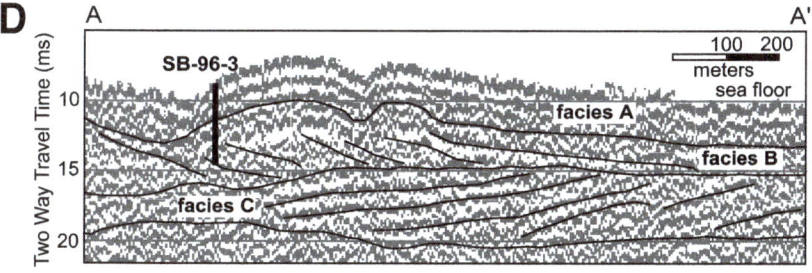

Figure 10. (A) Bathymetric map shows location of Sabine Bank study area (box). (B) Map of Sabine Bank shows sediment core locations and location of seismic line segment A–A′. (C) Representative lithological logs for sediment cores and associated sediment facies sampled by cores (designated with stars on B). Radiocarbon ages are also shown (in cal yr B.P.). (D) Interpreted seismic section across Sabine Bank shows seismic facies C (landward-dipping reflectors are interpreted as back-barrier bay and washover deposits), facies B (seaward-dipping reflectors are interpreted as shoreface deposits), and facies A (sub-parallel reflectors are interpreted as modern bank deposits). Modified from Rodriguez et al. (1999, 2004).

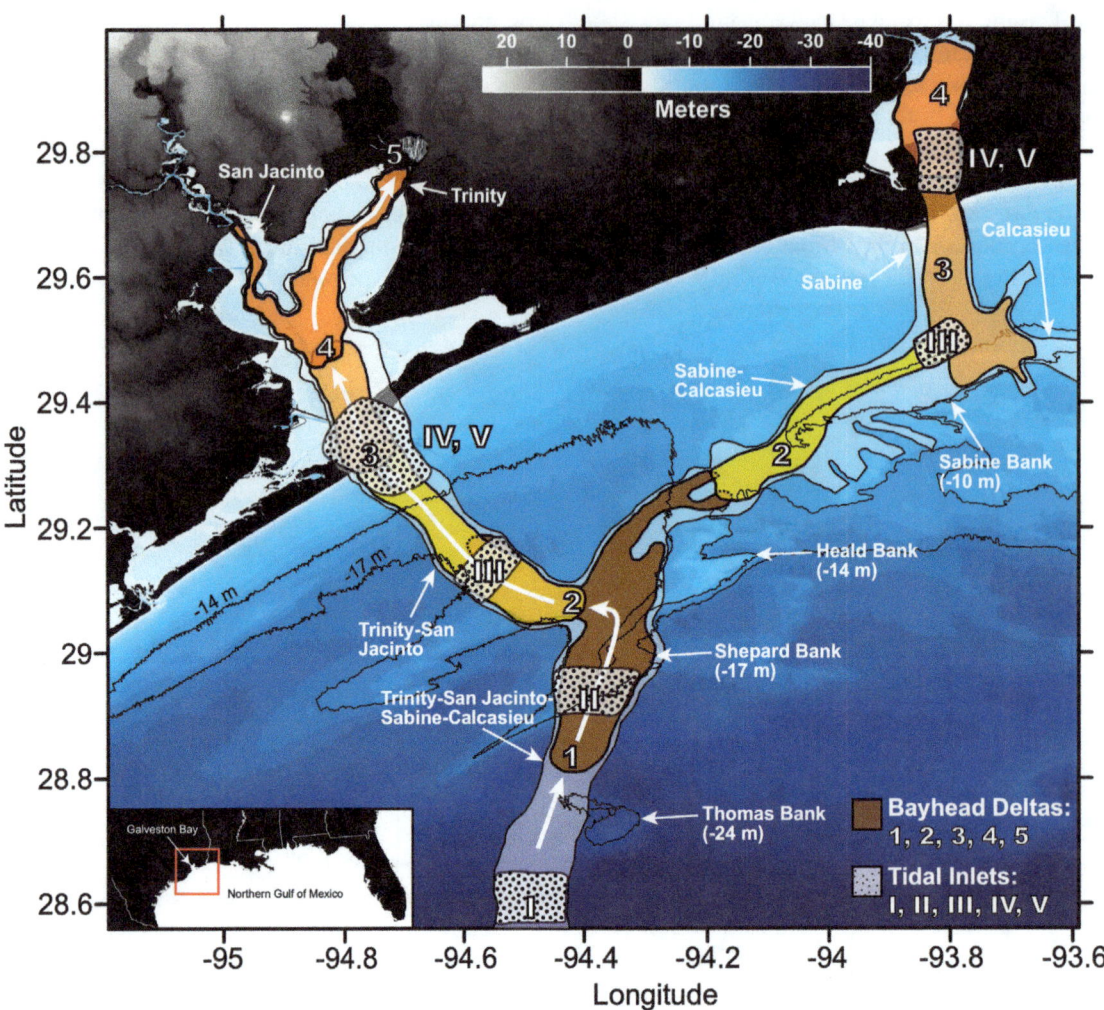

Figure 11. Map shows bayhead deltas and tidal inlets within the offshore Trinity and Sabine incised valleys with relative ages designated 1–5 for bayhead deltas and I–V for tidal inlets. Thicker white arrows indicate backstepping environments. Figure was modified from Thomas and Anderson (1994) and Anderson et al. (2016) to include additional Sabine valley results.

the earliest sea-level curves for the western Gulf also suggested episodes of rapid rise continuing into the early Holocene (Curray, 1960; Nelson and Bray, 1970), but these results are problematic due to large gaps in the data and imprecise dating.

Belknap and Kraft (1981) argued that incised fluvial valleys on continental shelves have high potential for preserving a record of transgressive coastal evolution because the sediments that fill these valleys may have escaped erosion during transgression, which in the study area occurs at water depths of up to −10 m (Rodriguez et al., 2001; Wallace et al., 2010). This argument prompted research that focused on late Pleistocene and Holocene strata that fill the Trinity and Sabine incised valleys aimed at learning more about the nature of early Holocene sea-level rise as it relates to coastal evolution (Thomas and Anderson, 1989, 1994).

The Trinity and Sabine incised valleys converge on the inner shelf ~50 km south of Bolivar Peninsula (Figs. 9 and 11). The stratigraphy of both valleys is characterized by stacked fluvial, bayhead delta, middle bay, and lower bay (tidal) facies that are bounded by prominent transgressive surfaces and capped by marine mud (Thomas and Anderson, 1989, 1994). These transgressive surfaces record landward shifts in estuarine environments that are typically greater than 10 km (Fig. 11). Anderson and Thomas (1991) argued that these landward shifts were caused by episodes of rapid sea-level rise due to rapid ice-sheet collapse. Their evidence for punctuated sea-level rise having caused these transgressive events was based on the observation that tidal delta, bay, and bayhead delta facies appear to be bounded by the same transgressive surfaces. Unfortunately, radiocarbon ages were not obtained, so the timing of these transgressive events was based on the locations and depths of the transgressive surfaces. These associations suggest that these flooding events occurred during the late Pleistocene and early Holocene.

Coastal Trinity, Sabine, and Calcasieu Valleys

Prompted by observations of apparently rapid landward shifts in bay environments within the offshore Trinity River and Sabine River valleys, studies were conducted in the onshore portions of the Trinity, Sabine, and Calcasieu incised valleys, which are now occupied by Galveston Bay, Sabine Lake, and Lake Calcasieu, respectively. This research was aimed at further defining the timing and magnitudes of the Holocene transgressive events (Anderson et al., 2008; Milliken et al., 2008b, 2008c). All three bays are characterized by stratigraphic successions composed of fluvial, estuarine, and marine facies that are typically bounded by discrete surfaces. To avoid the inference that these surfaces are necessarily caused by sea-level rise, we apply the term flooding surface. These three fluvial bay systems occupy similar climate and physiographic settings, which suggests that they would have similar flooding histories if the rates and magnitudes of sea-level events were sufficient to overprint differences in valley shape and sedimentation.

The methodology used to study all three bays consisted of first mapping the valleys using seismic data and collecting vibracores within modern bay environments to characterize their sedimentary and faunal facies. The final step involved collection of seismic lines along the axes of the valleys and then acquiring drill cores along these axial lines to examine each bay's stratigraphic evolution and obtain radiocarbon ages needed to constrain the timing of flooding events (Anderson et al., 2008; Milliken et al., 2008b, 2008c). These studies showed that estuarine environments within these bays are characterized by similar sedimentary and faunal facies and similar stratigraphic successions (Table 1). The deepest portions of the valleys contain fluvial sand, muddy sand, and gravel that was deposited during the MIS 2 lowstand and early transgression (Rehkemper, 1969). Above these fluvial deposits are bayhead delta deposits that include prodelta, mouth bar, and marsh (delta plain) facies that, when stacked in this order, indicate delta progradation. Where individual delta lobes were abandoned by lobe shifting, waves have reworked mouth bars to produce well-sorted sands with abundant *Rangia* shells (McEwen, 1969). The bayhead delta facies transitions offshore into middle bay facies. The lower bay is typically occupied by a flood tidal delta, which has a diagnostic seismic facies that includes obliquely dipping reflections marking lateral accretion and/or progradation into the bay and sediments comprised largely of tidal deposits including distinct tidal couplets (Table 1).

Galveston Bay. Galveston Bay occupies the incised river valley of the combined Trinity and San Jacinto rivers, which together have a modern sediment discharge of 6.2 10^6 t/yr and a combined drainage basin that occupies a wet, subhumid to humid climate zone (Fig. 3). The Galveston Bay Complex includes Trinity Bay, East Bay, and West Bay (Fig. 12), all with water depths that are on average less than 3 m deep.

Recognizing that the morphology of the valley would influence its flooding history, the first step in our investigation involved mapping the Trinity/San Jacinto incised valley. Initial work involved mapping the Pleistocene surface beneath Galveston Bay using an existing grid of U.S. Geological Survey (USGS) seismic (sparker) data (Smyth et al., 1988; Fig. 12A). This map

TABLE 1. BAY SEDIMENT AND FAUNAL FACIES

Sediments	Fauna
Fluvial	
Sand, muddy sand, and gravel	
Marsh	
Dark gray silty clays with abundant plant fragments	Rare arenaceous foraminifera and ostracods.
Mouth Bar	
Alternating fine sand and mud	Abundant *Rangia* shells and fragments. Foraminifera are common and include mostly *Ammonia beccarii* and, less commonly, *Elphidium* spp. and *Trochammina* sp.
Prodelta	
Gray clay with alternating thin organic layers	Macrofauna consists of *Crassostrea, Rangia, Mullinia,* and *Brachidontes.* Foraminifera are abundant and dominated by *Elphidium gunteri* and *Ammonia beccarii.*
Middle Bay	
Gray silty and sandy mud with abundant shell fragments	*Crassostrea, Rangia, Mullinia, Brachidontes,* and *Littoridina.* Foraminifera are dominated by *Elphidium gunteri* and *Ammonia beccarii.*
Lower Bay/Flood Tidal Delta	
Sand and interbedded sand and mud	Mixed bay and offshore species.

Sources: McEwen (1969); Rehkemper (1969); Wantland (1969); Poag (1981); Siringan and Anderson (1993).

shows a deep incision near the center of the valley and wide valley flanks with distinct terraces (Fig. 12B) that correlate to the established onshore middle Deweville, high Deweville, and Beaumont terraces (Thomas and Anderson, 1994). Both East Bay and West Bay are located outside the main valley, which is supported by vibracores that have sampled Pleistocene sediments beneath a few meters of Holocene bay sediments in both bays.

The next step in our Galveston Bay work was to acquire high-resolution (Uniboom) seismic data and vibracores from different bay environments to characterize their lithofacies and seismic facies (Anderson et al., 2008). An important framework for this work was provided by McEwen (1969) and Wantland (1969), who conducted detailed sedimentological and paleontological analyses of bay environments (Table 1). This was followed by acquisition of high-resolution seismic lines and a series of drill cores along the axis of the valley, which were needed to reconstruct the stratigraphic record of bay evolution (Fig. 12A). This phase of our research was aided by the work of Rehkemper (1969), who studied drill cores from the bay and provided an important stratigraphic framework for our seismic data interpretations. The combined results provide a record of bay evolution that is marked by landward stepping bay environments separated by discrete flooding surfaces that become less pronounced upward in the stratigraphic section (Anderson et al., 2008). A total of 32 radiocarbon ages were acquired from our drill cores that allow us to constrain the timing and duration of flooding surfaces and construct paleogeographic maps that illustrate the Holocene evolution of the bay (Fig. 13).

The deeper portion of the Trinity valley is filled with fluvial sands and gravels deposited during the MIS 2 lowstand and early transgression (Rehkemper, 1969). Initial flooding of the lower portion of the onshore valley occurred at ca. 9.6 ka (Fig. 13). This transgressive event resulted in an extensive bayhead delta that is correlated with Thomas and Anderson's (1994) offshore Delta 3 (Fig. 11). A flooding event at ca. 8.9 ka resulted in a landward shift of the bayhead delta and was followed by another transgressive event at ca. 8.2 ka that resulted in an open bay environment in the lower half of the bay and distributary channels of the bayhead delta being buried beneath prodelta and open bay deposits (Fig. 13A). This later flooding event was exacerbated as sea level

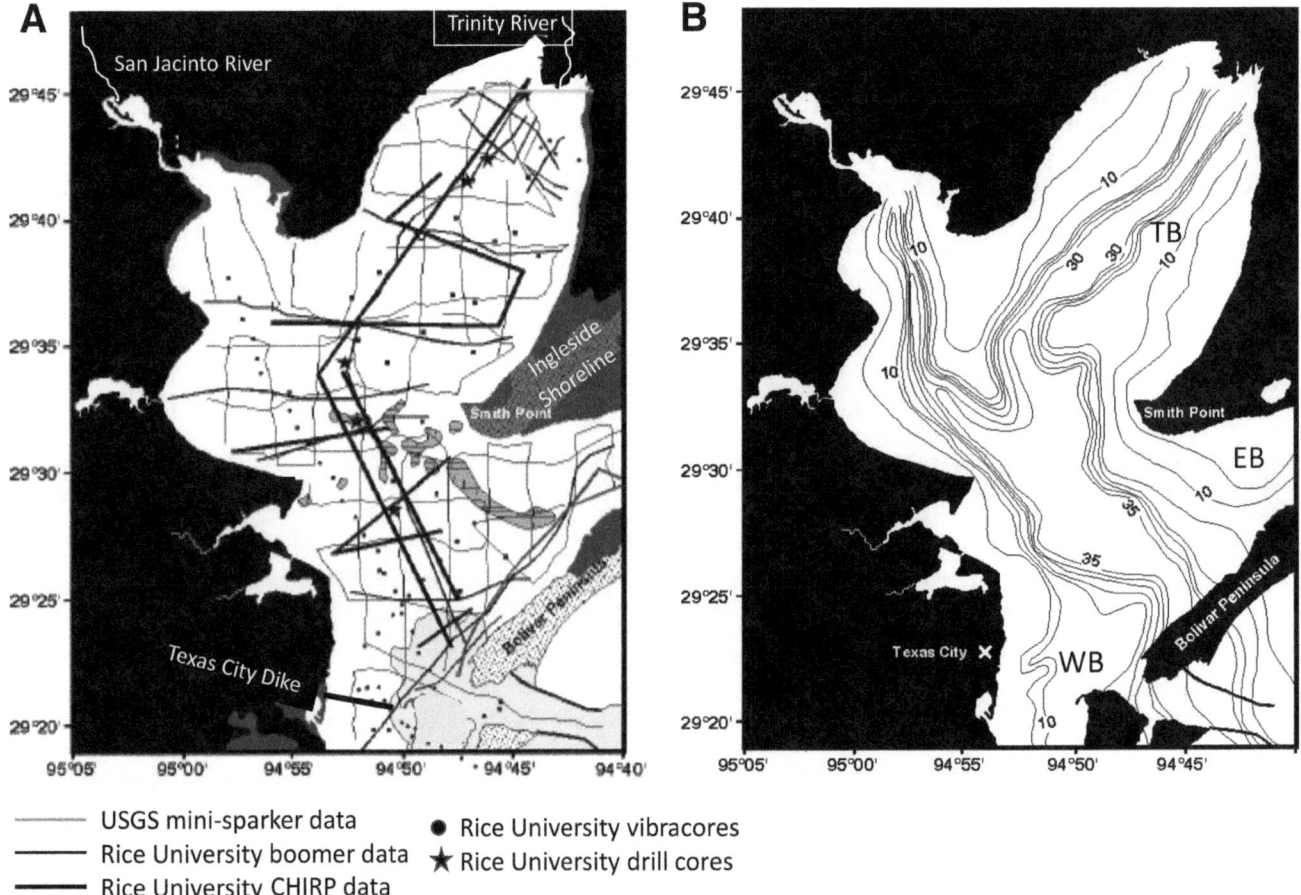

Figure 12. (A) Locations and sources of high-resolution seismic lines, sediment cores, and drill cores used to investigate Galveston Bay's evolution are shown. Boomer data, CHIRP data, and rotary drill cores are from Anderson et al. (2008). (B) Structure contour map of the Pleistocene surface beneath Galveston Bay. WB—West Bay, EB—East Bay, and TB—Trinity Bay. Figures were modified from Anderson et al. (2008).

reached the elevation of the relatively flat middle Deweyville terrace (Rodriguez et al., 2005; Anderson et al., 2008).

Following the ca. 8.2 ka flooding event, the delta front remained relatively fixed in its location until ca. 7.7 ka, when a major flooding event (7.7–7.4 ka flooding surface in Fig. 13A) resulted in the bayhead delta shifting northward ~30 km in ~300 years. Again, this flooding event was exacerbated by drowning of the low-relief high Deweyville terrace. During the ca. 8.2 ka and 7.7–7.4 ka flooding events, the Trinity bayhead delta was confined to the narrow and deep portion of the valley and retreated landward at an average rate of 6.5 km per century. It is notable that the early Holocene flooding events occurred when the average rate of sea-level rise was 4.2 mm/yr (Fig. 4B). Our results indicate that the average rate of sedimentation for the bayhead delta during this time was ~4.7 mm/yr, so the rates of sea-level rise and bayhead delta aggradation were similar (Rodriguez et al., 2010).

Following the 7.4 ka flooding event, the estuary setting within the Trinity valley changed relatively little as the rate of sea-level rise decreased to an average rate of 1.4 mm/yr (Fig. 4B). By ca. 4.0 ka, the average rate of sea-level rise decreased to 0.5 mm/yr, and the Trinity bayhead delta began to prograde and fill the upper portions of the bay, reaching its current location by 1670 cal yr B.P. (Fig. 13). By this time, both Bolivar Peninsula and Galveston Island existed and a large flood tidal delta extended nearly 20 km into the bay. As Bolivar Peninsula accreted toward the west, restricting the inlet between the peninsula and Galveston Island, this flood tidal delta decreased in size.

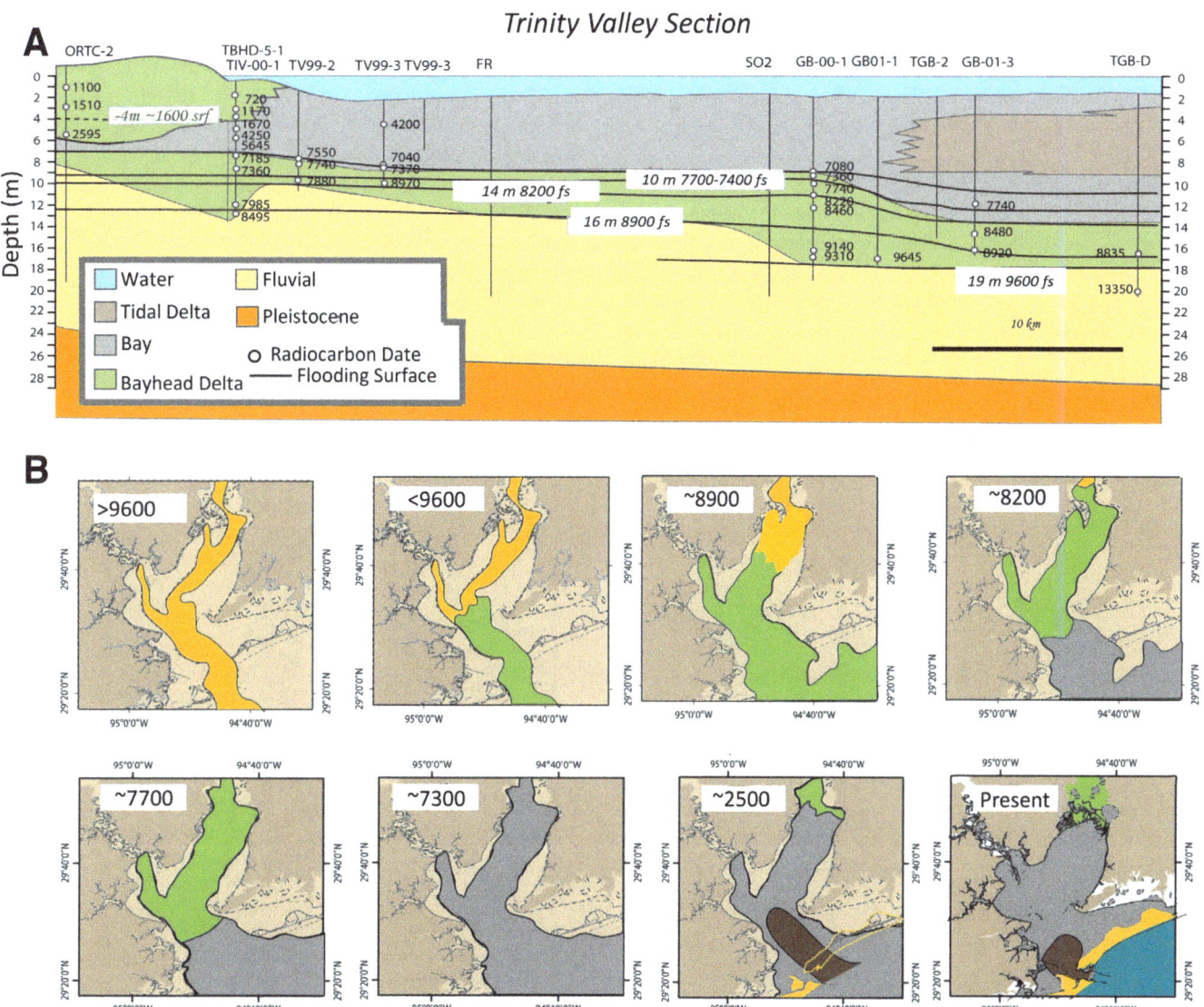

Figure 13. (A) North–south stratigraphic section for Galveston Bay was compiled from seismic data and drill cores collected along the axis of the Trinity River incised valley (see Fig. 12A). Also shown are flooding surfaces (fs) that mark episodes of landward-shifting bay facies; radiocarbon dates (in cal yr B.P.) are used to constrain the ages of these surfaces. (B) Paleogeographic maps depict changes in estuary evolution based on data shown in the above cross section and from other seismic and core data (modified from Anderson et al., 2008).

The modern Trinity Delta lobe prograded ~4 km during the past century (McEwen, 1969). This more recent phase of growth was possibly caused by deforestation and increased agricultural activity within the drainage basin (Phillips, 2010). The modern river mouth was stabilized in historical time, resulting in more focused growth of the delta as the western lobes experienced erosion. Al Mukaimi et al. (2018) observed that modern accumulation rates in the bay are currently not keeping pace with the rate of sea-level rise.

Sabine Lake. Two rivers, the Sabine River and the Neches River, drain into Sabine Lake. The two rivers have roughly equal drainage basin areas at 25,300 km² and 25,900 km², respectively, and a combined sediment discharge of ~3.7 10⁶ t/yr. Both drainage basins occur within the same humid climate zone as the lower Trinity River but with a somewhat higher average annual precipitation of between 100 cm/yr and 125 cm/yr (Fig. 3). Sabine Lake is a relatively shallow (–2.5 m on average) estuary that occurs within a broad coastal plain that is mostly less than 1 m above sea level. The development of the Sabine Chenier Plain has resulted in the formation of a narrow inlet between the estuary and the Gulf of Mexico (Fig. 14A).

Detailed study of Sabine Lake was conducted in the summer of 2003 (Milliken et al., 2008b) and included the acquisition of high-resolution seismic data, which, augmented by existing data sets, were used to map the Sabine/Neches incised valley (Fig. 14B). The approach and sampling rationale was the same as that used in Galveston Bay, with a seismic profile and sediment cores collected along the valley axis used to map seismic and lithofacies and identify bounding surfaces. In all, 11 vibracores and five drill cores were used to characterize and map the valley fill succession, and 32 radiocarbon ages obtained from shells and peat layers were used to constrain the ages of facies units and flooding surfaces (Fig. 15).

The Sabine/Neches valley exhibits an overall branching morphology that influenced valley flooding and evolution of the bay (Fig. 14B). The valley has terraces at –10 m to –12 m on its eastern side. Similar to Galveston Bay, the stratigraphic succession within the valley is characterized by a landward-stepping stratigraphic architecture composed of fluvial, bayhead delta, open bay, and tidal delta facies separated by flooding surfaces (Fig. 15A).

Initial flooding of Sabine valley to create a bayhead delta in the lower reaches of Sabine Lake occurred at ca. 9.8 ka. By ca. 8.9 ka, the bayhead delta shifted to the north ~10 km, marking a phase of northward expansion that would culminate at ca. 7.5 ka with the delta extending north of the study area (Fig. 15B). By ca. 7.1 ka, an open bay existed in the lower half of the estuary, including a flood tidal delta near the western margin of the bay. Between ca. 6.0 ka and ca. 5.5 ka, there was another significant flooding event that resulted in several kilometers of bayhead delta retreat and expansion of the flood tidal delta. Another flooding event at ca. 4.2 ka resulted in a landward shift in the flood tidal delta. After this, the estuary setting experienced little change until ca. 3.0 ka, when chenier development began to restrict the tidal inlet. By ca. 1.6 ka, the open bay had shifted north of the study area, and the eastern side of the flood tidal delta had aggraded to

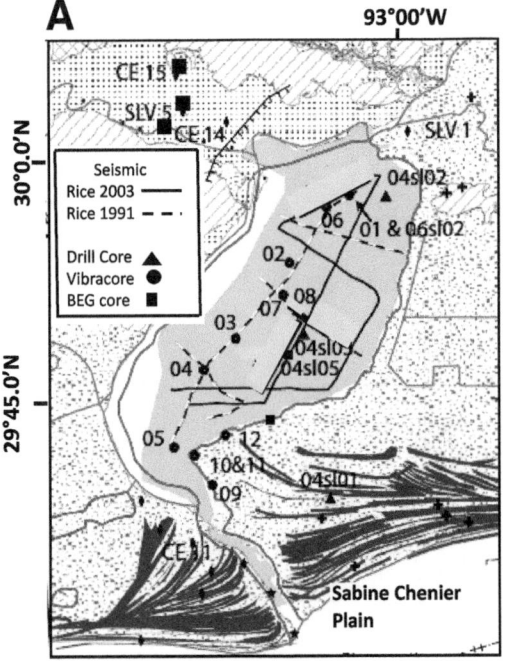
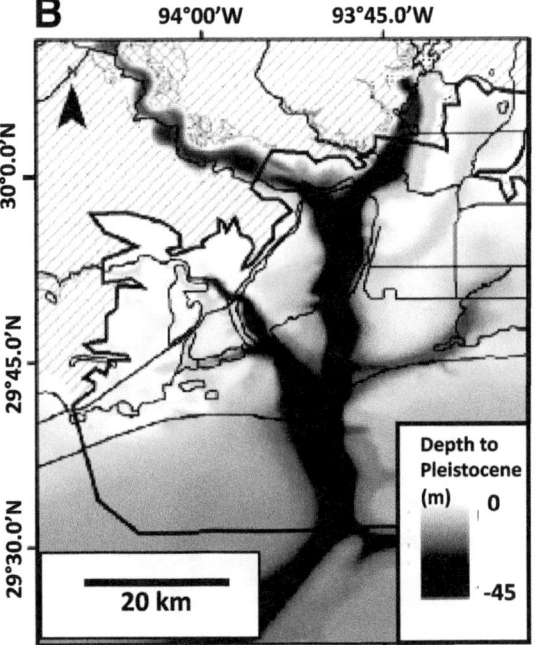

Figure 14. (A) Map shows locations of high-resolution seismic lines and sediment cores used to investigate the Holocene evolution of Sabine Lake. (B) Structure contour map of the Pleistocene surface beneath Sabine Lake. Figure was modified from Milliken et al. (2008b). BEG—Bureau of Economic Geology.

sea level with flood tidal delta deposits that extend ~8 km landward of Sabine Pass (Fig. 15B). Continued chenier plain development led to a more restricted connection to the Gulf and contraction of the flood tidal delta and mid-bay as the bayhead delta prograded to the south. As the flood tidal delta contracted, the ebb tidal delta expanded to become a prominent feature along the east Texas Coast.

Beginning ~200 years ago, the bayhead delta again began to retreat northward, and the flood tidal delta was buried in bay mud, giving the estuary its current shape and environmental setting. This phase of retreat occurred at the same time the Trinity bayhead delta was experiencing growth, which likely reflects differences in anthropogenic activity in the two drainage basins.

Lake Calcasieu. Lake Calcasieu has a semidiurnal to diurnal tidal cycle prism of 3.2×10^6 m^3, which flows through a narrow inlet to the Gulf of Mexico (Fig. 16A). The tidal range averages only 0.6 m, although the upstream tide penetration extends nearly 100 km (Forbes, 1988). This results in widespread dispersal of riverine sediments, the absence of a clearly defined bayhead delta, and a widespread open bay setting. As we will see, this

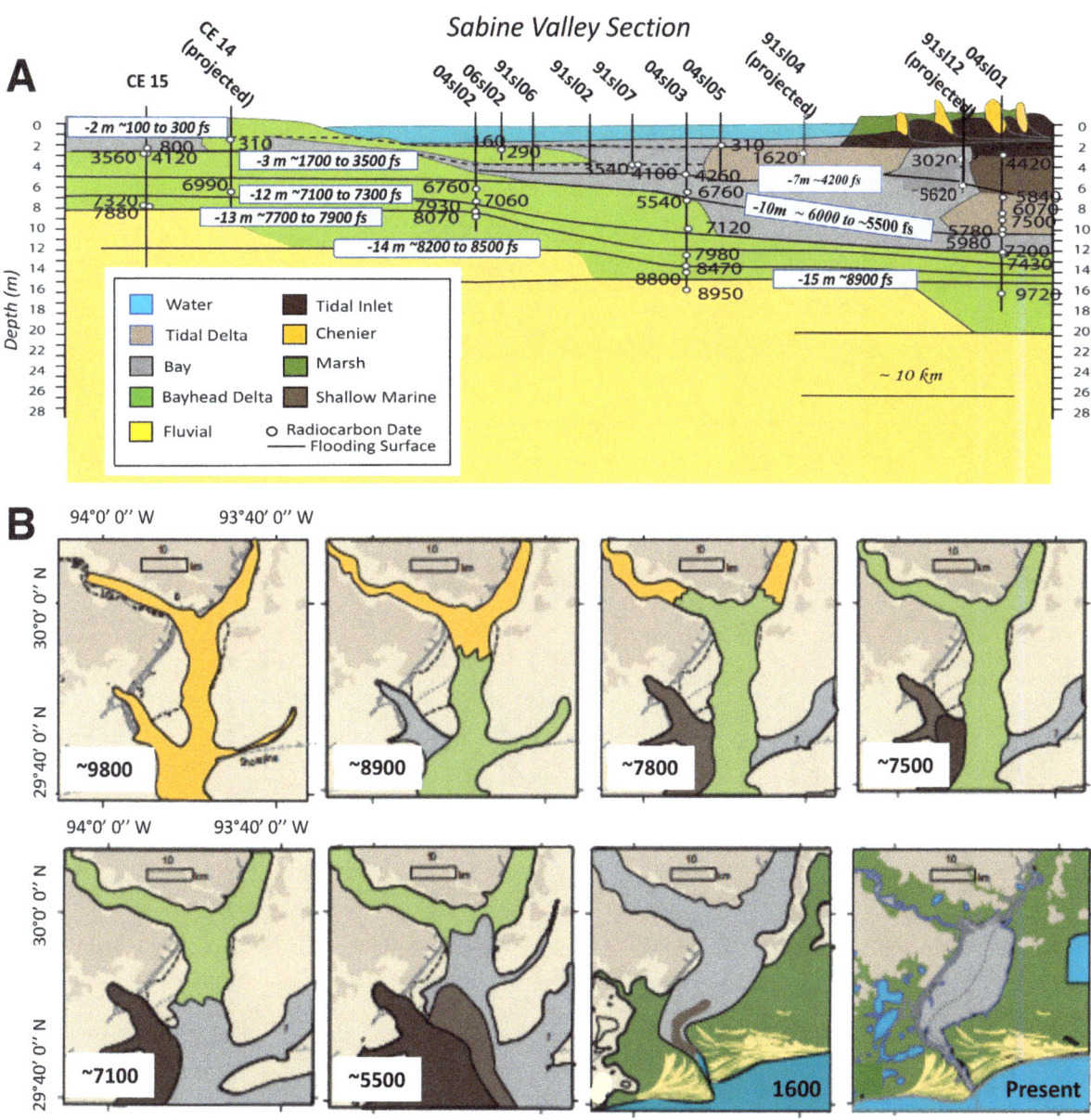

Figure 15. (A) North–south stratigraphic section for Sabine Lake was compiled from seismic data and drill cores (see Fig. 14A). Also shown are flooding surfaces (fs) that mark episodes of landward-shifting bay facies; radiocarbon dates (in cal yr B.P.) were used to constrain the ages of these surfaces. (B) Paleogeographic maps depict changes in estuary evolution based on data shown in the above cross section and other core data (modified from Milliken et al., 2008b).

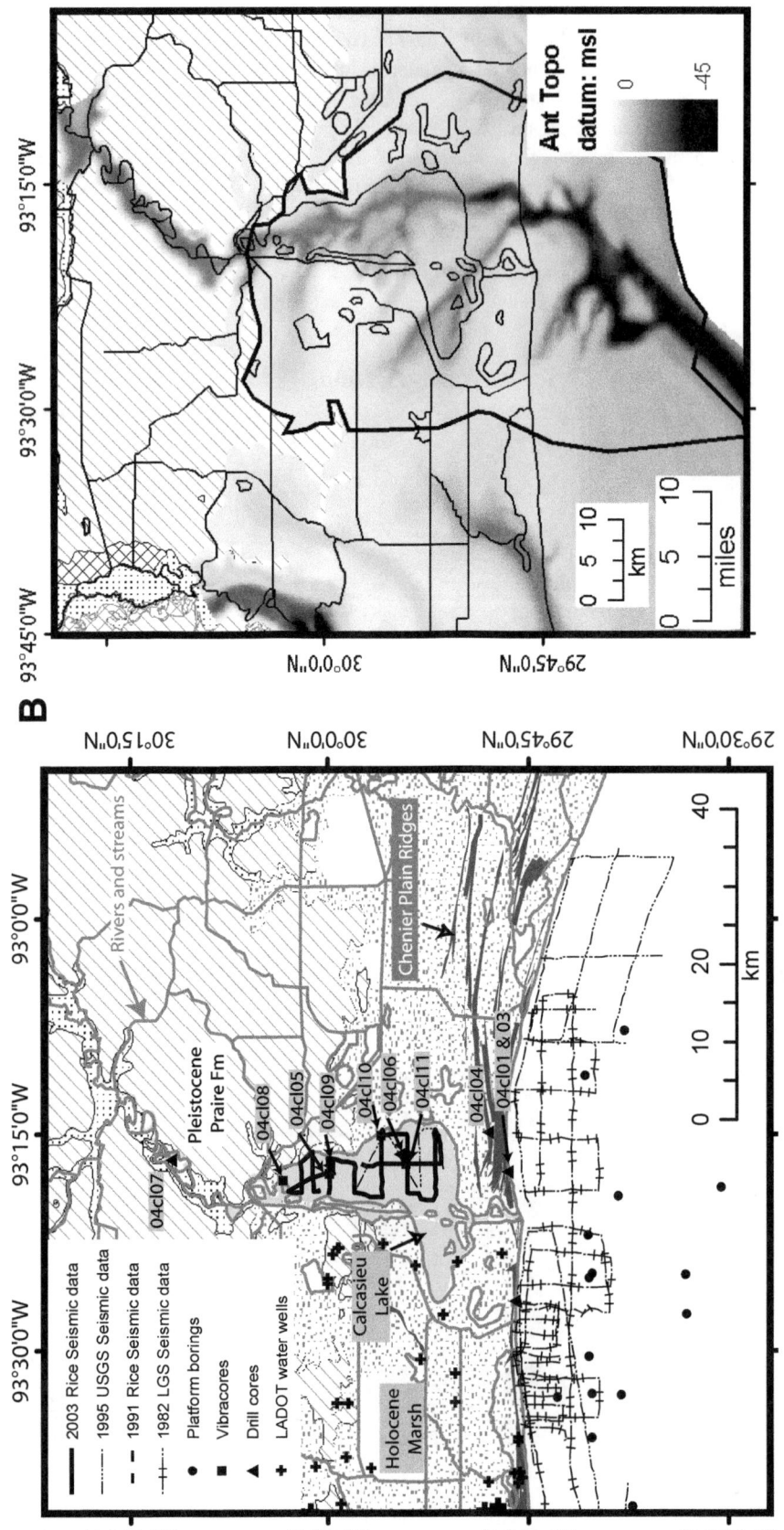

Figure 16. (A) Locations and sources of high-resolution seismic lines, sediment cores, and drill cores used to investigate Lake Calcasieu's evolution are shown. Note that some rivers and streams appear linear because they were dredged. (B) Structure contour map of the Pleistocene surface beneath Lake Calcasieu. Figure was modified from Milliken et al. (2008c). USGS—U.S. Geological Survey; LGS—Louisiana Geological Survey; LADOT—Louisiana Department of Transportation; Ant—Antecedent; msl—m above sea level.

condition is very different from that which occurred throughout most of the estuary's evolution.

Milliken et al. (2008c) acquired a grid of high-resolution seismic profiles within Calcasieu Lake that was used to refine previous maps of the Calcasieu River incised valley (e.g., Byrne et al., 1959; Nichol et al., 1996; Fig. 16A). Similar to Sabine Lake, the valley exhibits an overall dendritic pattern that had a strong influence on valley flooding and bay evolution (Fig. 16B).

Using the Nichol et al. (1996) map of the Calcasieu valley, Milliken et al. (2008c) focused on mapping seismic facies and flooding surfaces within the valley axis following the approach used in the Galveston Bay and Sabine Lake investigations. Eight drill cores and one vibracore were acquired along the axis of the valley to provide lithofacies information and material for radiocarbon dating of flooding surfaces (Fig. 17A). These results show that the evolution of Calcasieu Lake began in the early Holocene

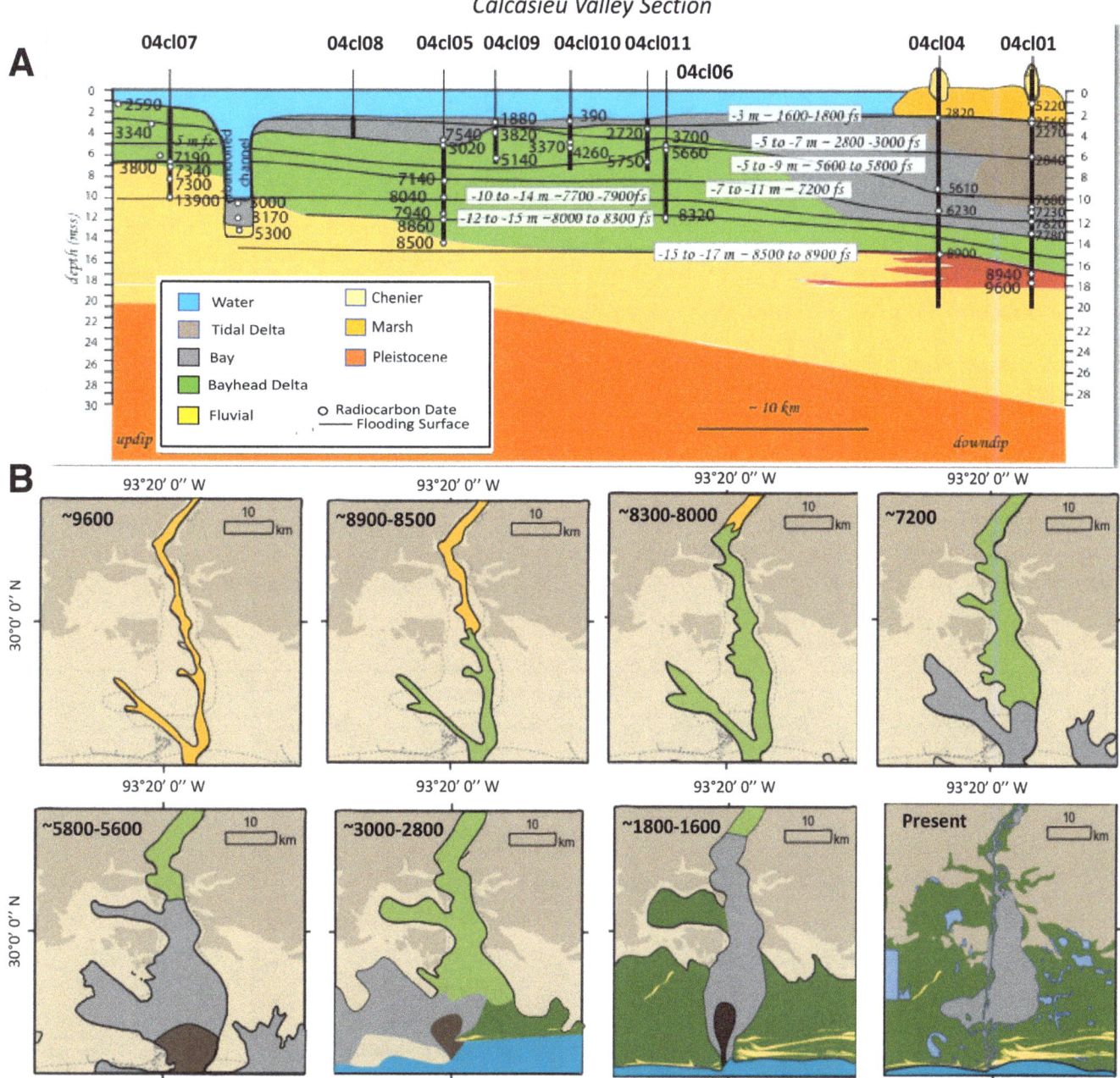

Figure 17. (A) North–south stratigraphic section for Lake Calcasieu was compiled from seismic data and drill cores (see Fig. 16A). Also shown are flooding surfaces (fs) that mark episodes of landward-shifting bay facies and radiocarbon dates (in cal yr B.P.) used to constrain the ages of these surfaces. (B) Paleogeographic maps depict changes in estuary evolution based on data shown in the above cross section and other core data (modified from Milliken et al., 2008c).

at ca. 8.9 ka, when a branching bayhead delta complex formed in the lower reaches of the estuary (Fig. 17B). Between ca. 8.3 ka to ca. 8.0 ka, a flooding event resulted in the bayhead delta retreating landward ~20 km. Open-bay conditions were established in the lower part of the bay by ca. 7.2 ka. This was followed by expansion of the open bay environment and establishment of a flood tidal delta between ca. 5.8 ka and ca. 5.6 ka. Between ca. 3.0 ka and ca. 2.8 ka, there was an expansion of the bayhead delta, which was followed by nearly 30 km of delta retreat. This change suggests a relatively short-lived increase in sediment supply to the estuary that abruptly ended at ca. 2.8 ka. The Calcasieu Chenier Plain began to form by ca. 2.8 ka and, similar to Sabine Lake, its growth led to restricted connectivity with the Gulf of Mexico, eventual demise of the flood tidal delta, and development of a prominent ebb tidal delta. By ca. 1.8 ka, the open bay setting extended north from the Chenier Plain to the vicinity of Lake Charles, Louisiana. Since then, the bayline has retreated south to its current location, and the estuary has taken on its current shape.

To summarize, there is strong evidence that the early Holocene evolution of Galveston Bay, Sabine Lake, and Lake Calcasieu was impacted by episodes of rapid sea-level rise. By mid-Holocene time, the role of sea-level rise in bay evolution decreased, and climate-controlled variations in sediment supply to these bays and coastal barrier/tidal inlet development played leading roles in their evolution.

Western Louisiana Chenier Plain

The western Louisiana Coast, from the western margin of the Mississippi River alluvial basin to Sabine Lake, is marked by an abrupt boundary between the Pleistocene fluvial plain to the north and Holocene coastal wetlands and lakes to the south. The modern Sabine and Calcasieu Chenier plains are prominent features that consist of alternating muddy swales and sandy ridges (Fig. 18). Narrow inlets separate Lake Calcasieu and Sabine Lake from the Gulf of Mexico, and a narrow strand plain separates the two cheniers. Early workers suggested that these unique coastal features were formed by alternating suspended sediment deposition and wave erosion of shorelines that was controlled by lobe shifting of the Mississippi Delta (Howe et al., 1935; Russell and Howe, 1935; Fisk, 1948; Gould and McFarlan, 1959). However, this theory was later discounted when it was discovered that major delta lobe shifting events had occurred over millennial time frames whereas mud/sand ridge couplets had formed over shorter timescales (McBride et al., 2007; Hijma et al., 2017).

Aerial photographs of both the Calcasieu and Sabine Chenier plains reveal that the sand ridges change from landward to offshore curvatures, which indicates relatively continuous growth and filling of tidal passes and a shift from flood-dominated to ebb-dominated tidal influence as these inlets became more restricted (Fig. 18). Radiocarbon ages indicate that the Sabine Chenier Plain began to form at ca. 3.0 ka to ca. 2.8 ka and has prograded up to 10 km since that time, yielding an average seaward growth rate of +3.3 m/yr. Optically stimulated luminescence (OSL) ages acquired by Hijma et al. (2017) indicate ~8 km of progradation (between ca. 2.9 ka and ca. 1.3 ka) for the Calcasieu Chenier Plain at an average rate of +5.0 m/yr.

Based on the age and the number of ridges in the Sabine Chenier Plain (~35), the frequency of ridge formation is estimated to be ~85 years, although the actual frequency may be quite variable. Regardless, this decadal formation frequency implies alterations in sediment supply from the Mississippi River operating at this frequency or formation by major storms. The former mechanism calls for periods when fine, suspended sediments from the

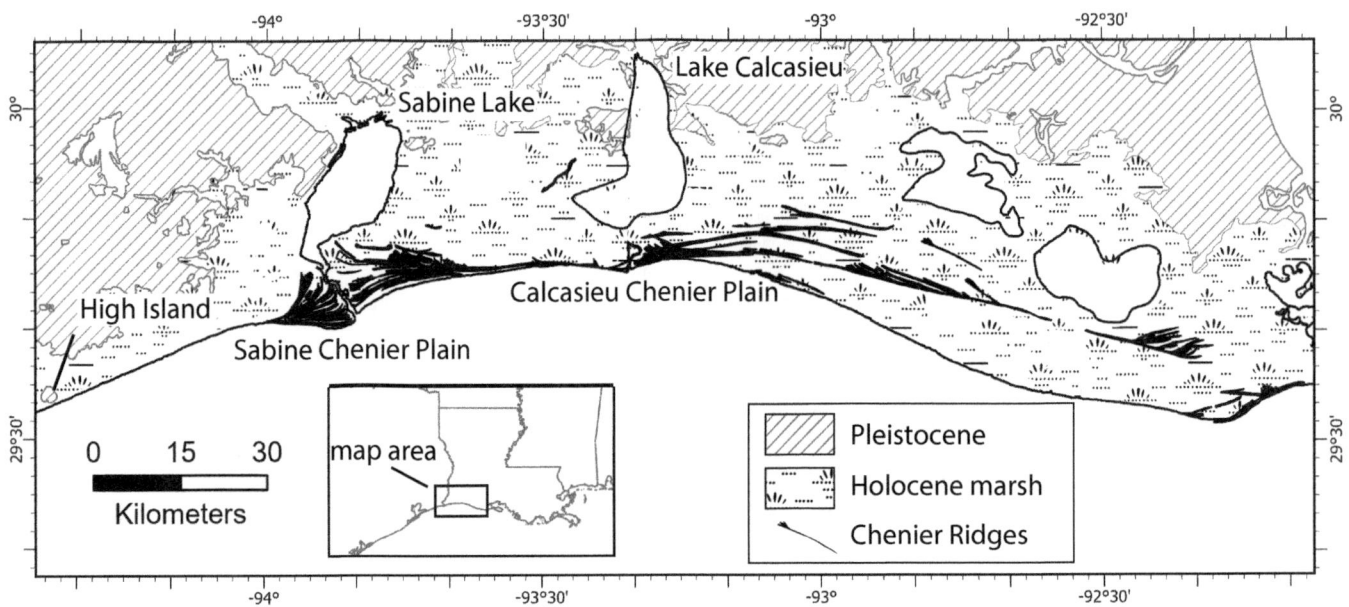

Figure 18. The western Louisiana Chenier Plain is shown (modified from McBride et al., 2007; © 2007 with permission from Elsevier).

Mississippi River overwhelmed sand supply to the coast, resulting in a mud-dominated coast followed by periods of decreased suspended sediment supply and wave reworking of the coast to form sand ridges. To our knowledge, there is no evidence for significant decadal variability in Mississippi River discharge to support the former mechanism for chenier formation. The latter mechanism calls for major storms eroding the cheniers to form sand ridges and is generally consistent with the frequency of major storm impacts in the area. This mechanism for chenier formation is further supported by a similar frequency for Bolivar beach ridge formation of ~83 years (nine ridges in ~750 years). Regardless of the exact formation mechanism, chenier development appears to have culminated in a recent phase of erosion that has removed portions of the younger, offshore-directed ridges.

East Texas Barrier System

The coast from Sabine Pass to High Island is a narrow beach that separates the Gulf of Mexico from an extensive, low-elevation coastal marsh (Fig. 18). Pleistocene sediments are exposed on the beach, and sediment cores from the shoreface and inner shelf sampled Pleistocene deposits at less than 1 m below the seafloor (Williams et al., 1979; Siringan and Anderson, 1994). Historical shoreline migration rates along this stretch of coast average −2.5 m/yr (Paine et al., 2012).

Bolivar Peninsula. Bolivar Peninsula is believed to have nucleated from a spit that was located slightly seaward of its current location (Morton, 1994). Detailed research by Siringan and Anderson (1994) and Rodriguez et al. (2004) relied on aerial photographs and LIDAR images, high-resolution seismic profiles, multiple vibracores, and observations within sand pits on the peninsula to examine the history of the barrier (Fig. 19). This research included the acquisition of 40 radiocarbon ages that were used to constrain the timing of barrier formation.

The western portion of Bolivar Peninsula is situated over the Trinity River incised valley, which has been largely filled with fluvial, estuarine, and tidal deposits (Figs. 12B and 20A). Barrier sands rest on an irregular tidal erosion surface that slopes toward the west. Combined seismic data, sediment cores, and engineering platform borings show that the thickness of the barrier increases from ~1.5 m at its east end to nearly 10 m at its west end. Initial progradation and lateral growth of the peninsula likely occurred rapidly as sand supply filled the shallow accommodation along the eastern margin of the Trinity valley and decreased as the peninsula migrated to the west over the deeper portions

Figure 19. Locations of sand pits (Q1–Q4 designated with squares) and sediment cores (white and red circles) used to study the Holocene evolution of Bolivar Peninsula are shown. The base map is a digital elevation map based on LIDAR data from the National Oceanic and Atmospheric Administration. Also shown are beach ridge sets used to locate cores. Ridge sets 1 and 2 are Holocene features and ridge set 3 formed in historical time. Red squares designate sand pits studied in detail and shown in Figure 21, and red circles are sediment cores shown in Figure 22.

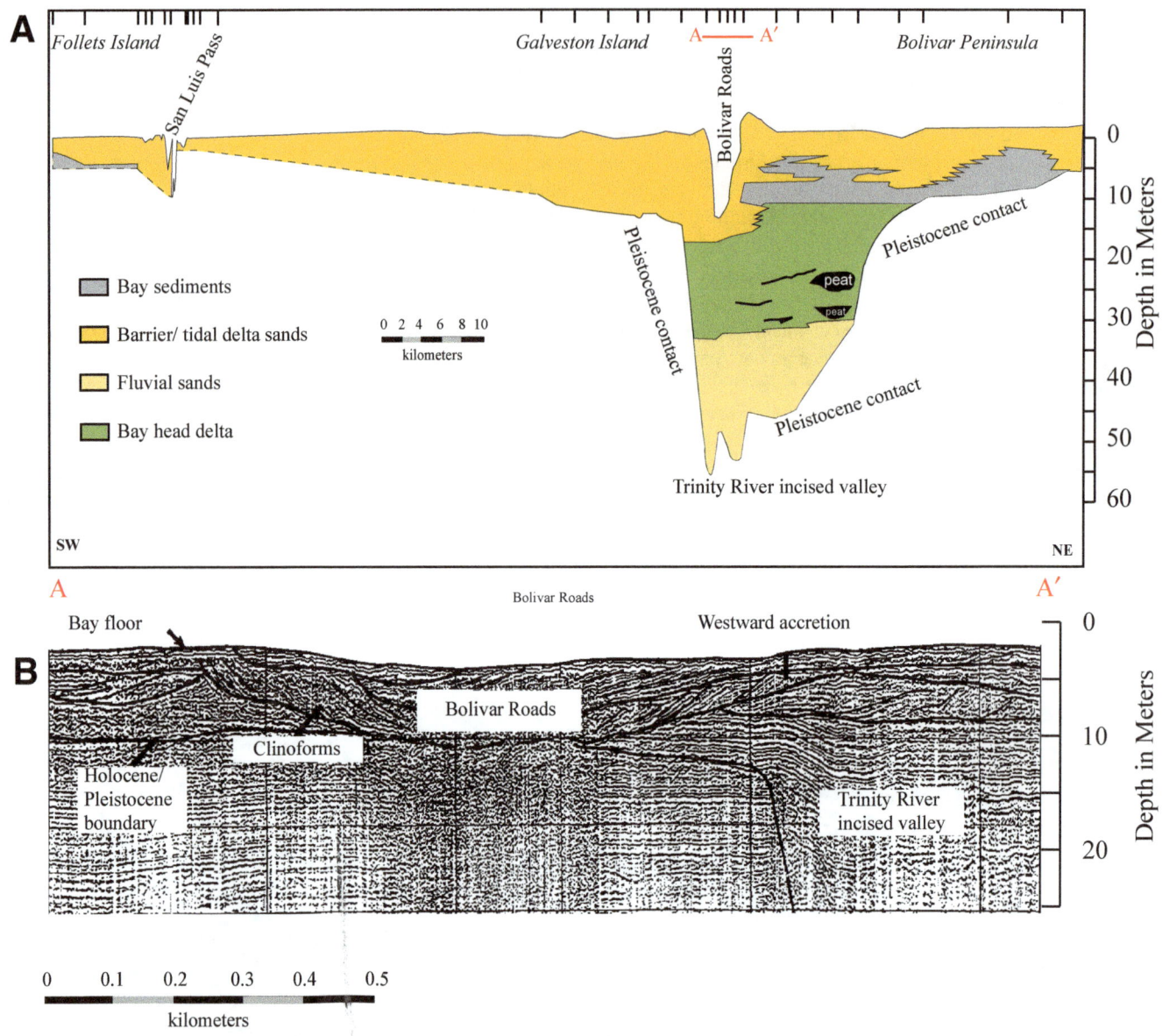

Figure 20. (A) Shore-parallel stratigraphic section through Bolivar Peninsula to the east end of Follets Island is shown (modified from Wallace et al., 2009). Black vertical lines at the top indicate core locations. Also shown is the seismic profile A–A' location. (B) High-resolution seismic profile across Bolivar Roads tidal inlet shows large-scale cross bedding and clinoforms that formed as the inlet became narrower, mainly by westward accretion of Bolivar Peninsula (modified from Siringan and Anderson, 1993).

of the valley. Seismic data and sediment cores from the Bolivar Flood Tidal Delta show a history of westward migration followed by inlet stability and a decrease in the size of the delta (Fig. 20B; Siringan and Anderson, 1993).

Aerial photographs and LIDAR images show three beach ridge sets on Bolivar Peninsula (Fig. 19). The oldest set extends along the north side of the peninsula and curves bayward along the western portion of the peninsula, recording westward migration of the tidal inlet and accretion of the peninsula. These older ridges are overprinted by storm washover features, including two prominent washover fans. A younger set of beach ridges increases in width toward the west where they curve bayward, recording a phase of seaward progradation and westward accretion of the western portion of the peninsula. The youngest set of ridges extends ~10 km along the western end of the island. Historical charts show that this youngest beach ridge set formed after rock jetties were constructed on either side of the Galveston-Houston Ship Channel in the late 1800s.

The overall beach ridge morphology of Bolivar Peninsula indicates erosion and storm overwash on the east end of the peninsula, and progradation and lateral accretion on the western part of the peninsula. The beach ridge morphology served as a guide for selecting core transects across the peninsula and acquiring radiocarbon ages needed to investigate its evolution (Rodriguez et al., 2004). Experience has shown that shells obtained from beach ridges are often reworked and yield anomalously old ages. To the degree possible, we used *Donax* shells, a species that is known to inhabit the swash zone and proximal shoreface environment, for radiocarbon dating. Articulated *Donax* shells or those with both halves intact, which are rare, and shells found in matching pairs but not intact are particularly reliable for dating. We also sampled shells from washover fans that represent proximal bayline settings.

Prior to the late 1990s, sand mining was allowed on Bolivar Peninsula, with four large sand quarries providing exposures near the middle of the island and within the older beach ridge field (Fig. 19). All four sand quarries expose barrier sands separated from underlying washover deposits, bay mud, and organic-rich marsh deposits by a prominent shell lag (Fig. 21). We were able to access and sample three of these pits for more detailed analysis. The barrier sand ranges from 1.0 m thick to 1.5 m thick and consists of well sorted, fine sand with discrete shell beds near the base of the section. *Donax*, *Mulinia*, *Anadara*, and reworked *Crassostrea* are the most common mollusks found within the barrier sands. The shell bed at the base of the barrier sand unit contains a mixed assemblage of bay mollusks, mostly *Crassostrea* and *Rangia*, and offshore species. *Rangia*, *Mulinia*, and *Crassostrea* shells from this shell bed yielded a broad range of ages that are not in stratigraphic order, so the youngest ages are used to establish the maximum age of the shell beds in Quarries

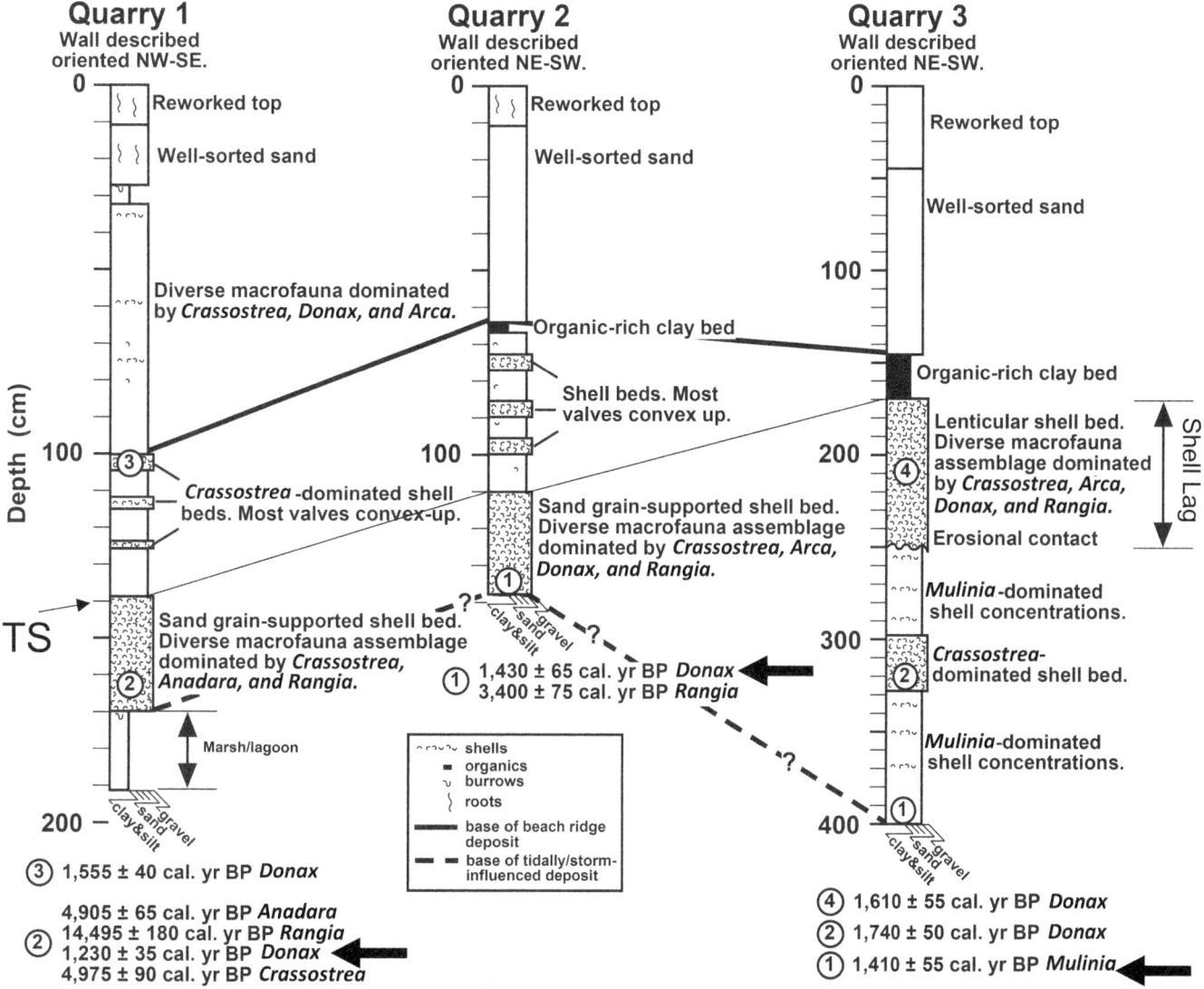

Figure 21. Stratigraphic sections and radiocarbon ages for three Bolivar Peninsula sand quarries are shown (modified from Rodriguez et al., 2004). These quarries sampled two Holocene beach ridge sets overlying an organic-rich marsh unit that rests sharply on a shell lag. The shell lag is interpreted as a transgressive deposit and the bounding surface as a transgressive surface (TS). The shell lag overlies bay mud. See Figure 19 for pit locations.

1 and 2. Radiocarbon ages from mixed *Donax* shells from the barrier sands in Quarry 1 yielded an age of 1555 cal yr B.P. The youngest ages from the shell beds in all three quarries range from 1230 cal yr B.P. to 1430 cal yr B.P. An articulated *Mulinia* shell from bay mud beneath the shell bed in Quarry 3 yielded an age of 1410 cal yr B.P. These results indicate that the peninsula was at its current location and experienced a major transgressive event at ca. 1.4 ka that lasted until after 1.2 ka.

Beach ridge morphology indicates that the peninsula was breached on two occasions, which led to the formation of two prominent washover fans (Fig. 19). Both fans have beach ridges that curve landward near the breaches, which indicates that these features were active tidal inlets for some length of time. Sediment cores were collected from these fans to determine their age, recognizing that washover fans and tidal deltas commonly contain reworked shell material and that multiple dates are required to constrain their age. The youngest age acquired from washover deposits of the eastern fan was 1530 cal yr B.P., and the youngest age from washover deposits of the western fan was 1755 cal yr B.P. These ages are slightly older than the ages from the sand pits indicating that the peninsula has been located near its current location since 1755 cal yr B.P. but was initially narrow enough to experience breaching and washover. There is no indication that the barrier extended much further landward, as core EB-02-1 from East Bay sampled bay mud with ages that are in stratigraphic order and are as old as 7720 cal yr B.P. (Rodriguez et al., 2004).

Two core transects were acquired across the peninsula, one near its center and the other near the western end (Figs. 19 and 22). Onshore cores from the transect near the center of the peninsula penetrated barrier sand resting on proximal washover and bay deposits (Fig. 22A). The most landward core in the central transect (BP-00-10) sampled 25 cm of proximal washover sands resting on bay mud. Core BP-00-11 sampled a succession of bay mud overlain by distal washover deposits with proximal washover sand at the top. Cores BP-00-12 through BP-00-14 were collected near the gulf side of the peninsula and sampled 180 cm and 250 cm of barrier sand with shell hash layers near the base of the sand units. Core BP-00-12 yielded two radiocarbon ages

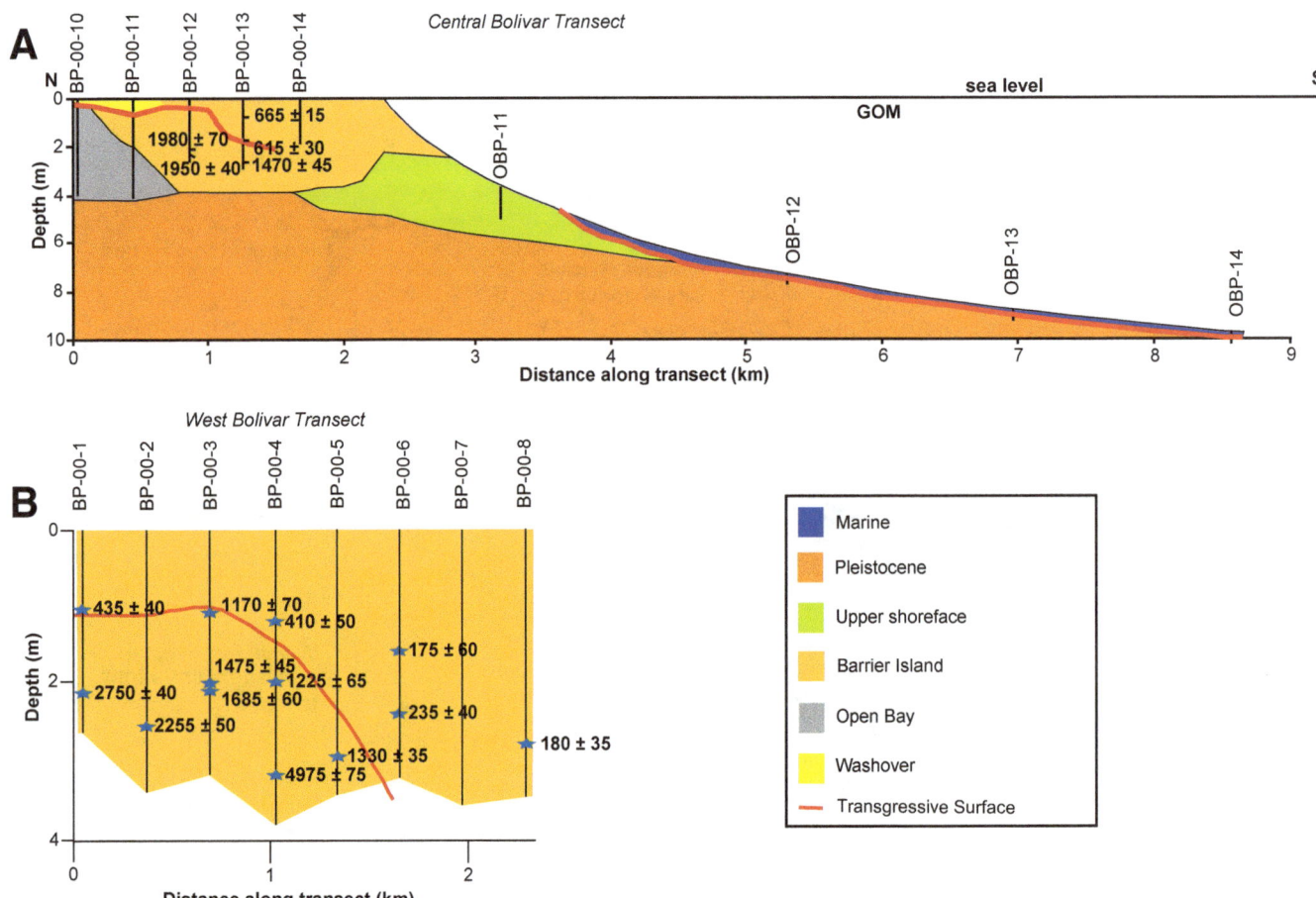

Figure 22. (A) Onshore to offshore central Bolivar Peninsula stratigraphic section is shown. Core depths and water depths are not exactly to scale. (B) Onshore stratigraphic section for west Bolivar Peninsula. Both transects sampled a transgressive surface that separates beach ridges older than 1170 cal yr B.P. from ridges that are younger than 650 cal yr B.P. Offshore, this transgressive surface is marked by marine mud that onlaps shoreface deposits. See Figure 19 for core locations. Figure adapted from Rodriguez et al. (2004). GOM—Gulf of Mexico.

near the base of the barrier sand unit of 1980 cal yr B.P. and 1950 cal yr B.P., which indicate that the peninsula was at this location at that time.

The two youngest beach ridge sets were sampled in core BP-00-13, which was collected ~200 m south of Core BP-00-12. This core sampled a lower unit of barrier sand with shell hash beds and an upper unit composed of two well sorted sands that were interpreted as beach ridge deposits separated by an organic-rich clay that is interpreted as a swale deposit. A sample from the lower sand unit yielded a radiocarbon age of 1470 cal yr B.P. The upper sand unit yielded two radiocarbon ages of 615 cal yr B.P. and 665 cal yr B.P. Thus, this core sampled the erosional surface that separates the two oldest beach ridge sets and indicates that this surface cuts deep into the peninsula (Fig. 22A).

Offshore cores sampled a thin layer of upper shoreface sands that extend just over 1 km to ~6 m water depth (Fig. 22A). Seaward of this (~1.5 km from the modern shoreline), cores sampled marine mud resting directly on Pleistocene clay. Thus, a relatively shallow transgressive surface occurs at ~6 m water depth. We assume that this surface correlates to the transgressive surface that separates the older and younger beach ridge sets, but no radiocarbon ages are available from the offshore cores.

The western core transect was intended to sample all three sets of beach ridges (Fig. 19). Radiocarbon ages from these cores indicate that the older beach ridge set is buried beneath 1 m of sand in our most landward core BP-01 (Fig. 22B). The older ages are not all in stratigraphic order, but if only the youngest ages are considered, the cores show a general trend of decreasing age in a seaward direction and indicate barrier progradation between 2750 cal yr B.P. and 1170 cal yr B.P. One age of 4975 cal yr B.P. from the bottom of core BP-00-4 is believed to be reworked. As observed in the cores from the central peninsula transect, there is an erosional surface that separates the older episode of progradation from the younger set of ridges, which yielded ages younger than 435 cal yr B.P.

In summary, the radiocarbon ages from the stratigraphic successions observed in sand quarries and in vibracores from Bolivar Peninsula indicate that the barrier has been located near its current location since 2750 cal yr B.P. and experienced lateral and vertical growth until at least 1170 cal yr B.P. Core transects from the central and western parts of the peninsula sampled an unconformity that separates these older barrier deposits from barrier sands associated with prominent beach ridges that yielded radiocarbon ages younger than 665 cal yr B.P. This unconformity marks a major event that severely eroded the peninsula. It is interpreted to have been caused by a single intense storm or to have developed during a stormy period. This event was followed by a phase of healing and shoreline progradation that lasted several centuries (Rodriguez et al., 2004). Since the 1867 chart of the inlet was drafted, up to 2 km of progradation has occurred near the western end of the peninsula where longshore sand transport is being blocked by the north jetty of the Galveston-Houston Ship Channel (Fig. 19). East of this area, beach ridges are truncated at the modern shoreline, and the average historical shoreline erosion rate is −1.3 m/yr (Paine et al., 2021). So, with the exception of its far west end, the peninsula has shifted from net growth to net erosion in historical time.

Galveston Island. Galveston Island has been the focus of research for more than six decades, and its morphology, sedimentology, and history have been well studied (Bernard et al.,1959, 1970; Rodriguez et al., 2004). With a maximum width of 6 km at its eastern end, the island tapers to the west and decreases to 1.3 km wide at its western end (Fig. 23). The thickness of the barrier sand also decreases from 12 m to 2 m from east to west, which reflects the relief on the underlying Pleistocene surface associated with the ancestral Trinity River incised valley that was subsequently modified by tidal erosion (Fig. 20). Two discrete sets of beach ridges record a history of seaward progradation of the island, except for along its western portion, where washover features dominate the landscape. An older, more landward set of ridges trends toward the bay near the east end of the island and is breached by storm surge channels (Fig. 23C). A younger, more continuous seaward set of ridges is oriented parallel to the current shoreline and shows little evidence of breaching. There are no well-developed subaerial washover fans on the eastern end of the island, although storm surge channels and associated washover deposits dominate the backshore landscape of the western part of the island.

The age of beach ridges and timing of seaward growth was partially constrained by Bernard et al. (1959, 1970) using a transect of drill cores through the island at 8-Mile Road (Figs. 23C and 24). Shells collected from the oldest beach ridges near the landward side of the island yielded radiocarbon ages as old as ca. 5.4 ka, and samples collected in more seaward locations yielded generally younger ages, although many of the radiocarbon ages acquired by these early workers were not in stratigraphic order and were bulk samples that included shells that were likely reworked. Rodriguez et al. (2004) acquired additional sediment cores along a transect near 8-Mile Road and used *Donax* shells and organic-rich swale deposits to constrain barrier evolution. With few exceptions, these ages are in stratigraphic order and are generally consistent with ages from Bernard et al. (1970) (Fig. 24). These results indicate a relatively slow rate of barrier progradation of 0.4 m/yr between 5300 cal yr B.P. and 2600 cal yr B.P. This overlaps with ages from Sabine Bank and is consistent with a large embayment in the coastline over the Trinity River incised valley at this time (Rodriguez et al., 2004). Results from heavy mineral analyses of sand samples indicate a shift from Trinity River–derived sand to sand that was reworked from offshore sources at about the time the beach ridge orientation changed (Cole and Anderson, 1982). Thus, early development of the island was slow as it struggled to fill the Trinity River valley. This was followed by more rapid progradation (1.25 m/yr) after ca. 2600 cal yr B.P., which is about the same time that Bolivar Peninsula began to experience relatively rapid progradation. Rodriguez et al. (2004) argued that progradation of Galveston Island ended sometime after ca. 1800 cal yr B.P., the age of the youngest beach ridge on the island. To better constrain when Galveston Island

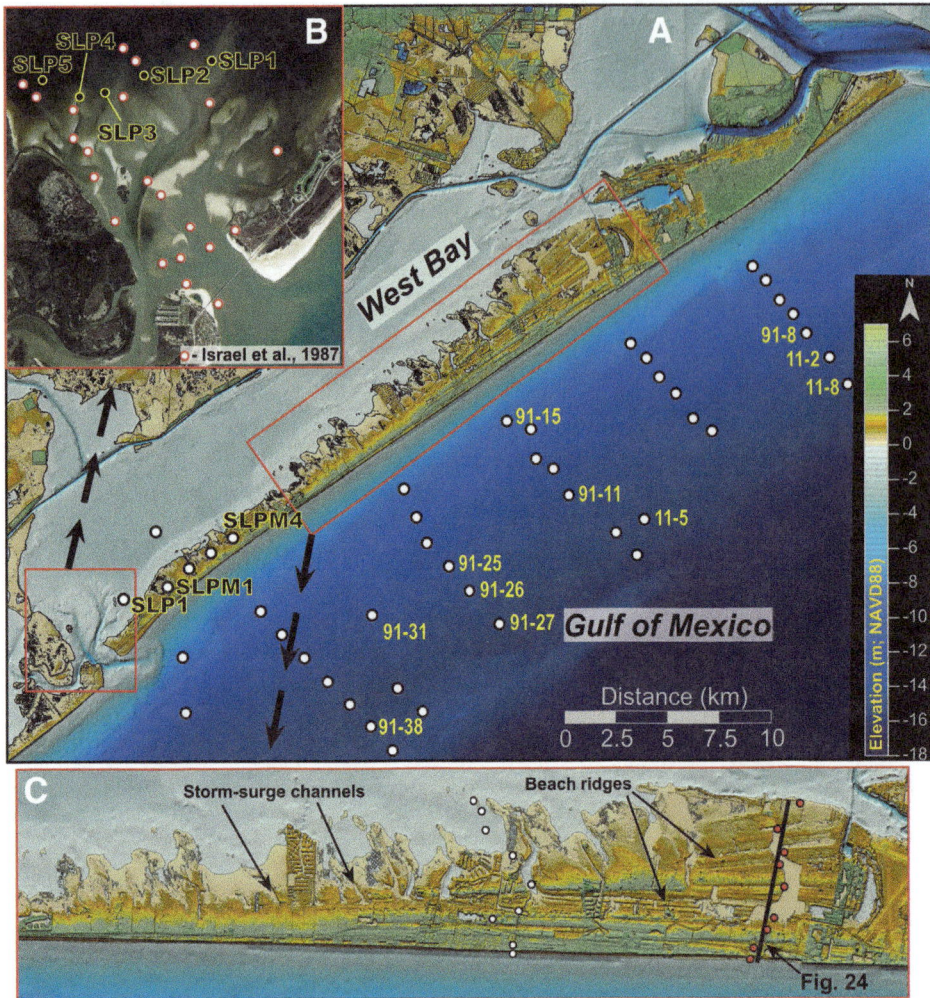

Figure 23. (A) Sediment cores used to investigate Galveston Island's evolution. Red boxes show locations of panels B and C. (B) The inset aerial photograph shows core locations for the San Luis Pass (SLP) Flood Tidal Delta. (C) Digital elevation map constructed from LIDAR data (from the National Oceanic and Atmospheric Administration) shows two sets of beach ridges, an older set that is oriented oblique to the coast and a younger, shore-parallel set. Storm surge channels on the east end of the island breach only older beach ridges, whereas they dissect the island farther to the west where the barrier is narrower and lower in elevation. Also shown are locations of sediment cores used to construct Figure 24 (red dots). White dots in panel C indicate additional core locations.

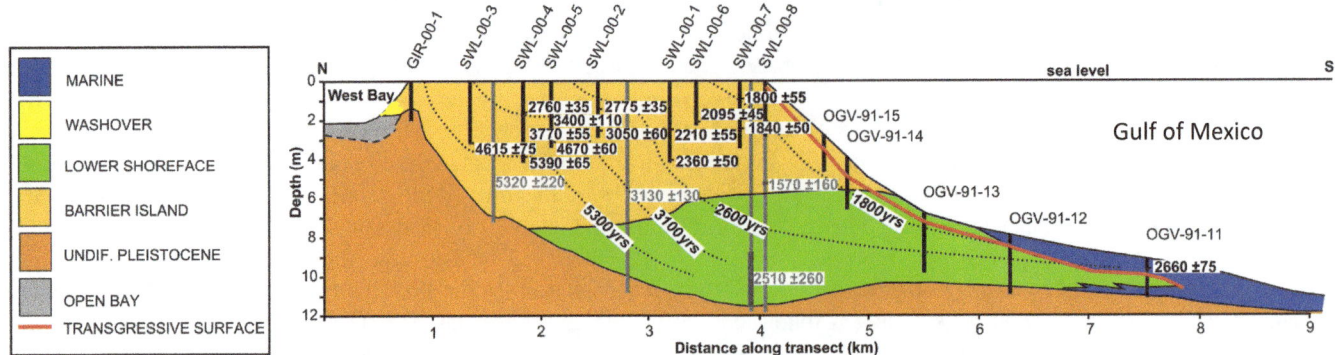

Figure 24. Cross section through Galveston Island at 8-Mile Road is based on lithologic descriptions and radiocarbon ages from Bernard et al. (1970) and vibracores (labeled GIR, SWL, and OGV). Onshore cores were collected and studied by Rodriguez et al. (2004) and yielded radiocarbon ages that are generally consistent with those of Bernard et al. (1970) and indicate seaward progradation of the island spanning ca. 5.3 ka to ca. 1.8 ka (dashed lines are isochrons). The offshore cores were acquired by Siringan and Anderson (1994) and record shoreface progradation to about the location of core OGV91-11, followed by onlap of marine mud since ca. 2.7 ka, which marks the reversal from progradation to retrogradation.

shifted from a progradational mode to a transgressive mode, we turn to the offshore record.

Siringan and Anderson (1994) used five upper shoreface to inner shelf core transects to examine the stratigraphic record of barrier migration (Fig. 23A). Transect 3 collected approximately 10 kilometers west of the 8-Mile Road transect is used to illustrate the final stages of barrier progradation followed by transgression (Fig. 24). The progradational succession is composed of shoreface deposits that extend nearly 4.0 km offshore. The transgressive succession is marked by ~1.5 km onlap of marine mud onto shoreface deposits. The more distal cores in all five transects sampled Pleistocene deposits at a water depth of ~10 m, which indicates that this is the maximum depth of transgressive erosion. Core OGV91-11 records the most seaward extent of lower shoreface deposits and yielded an age of 2660 cal yr B.P. from marine muds resting above distal shoreface deposits (Fig. 24), which marks the onset of marine transgression at this location (Rodriguez et al., 2004).

Wallace and Anderson (2013) later acquired additional offshore cores and re-occupied some of the core sites of Siringan and Anderson (1994) in an effort to better constrain volumetric erosion and to examine the long-term offshore sand flux of the island. They collected 16 new offshore radiocarbon ages, all of which were in stratigraphic order but showed spatial variability. During the early part of Galveston Island's history (ca. 5100 cal yr B.P. to present), the average offshore storm sand flux was ~4300 ± 670 m^3/yr. From ca. 2700 cal yr B.P. to present, the offshore sand flux dramatically decreased to ~950 ± 270 m^3/yr, which is largely a function of the aforementioned landward retreat of the island due to reduced sediment supply at ca. 1800 cal yr B.P. This declining flux suggests that hurricane impacts have been minor yet constant contributors to island erosion over millennial timescales. If we assume that the shoreface and barrier were in full retreat by 1.8 ka and that the lower shoreface has retreated ~1.3 km since that time to present, as indicated by the extent of marine mud onlap of shoreface deposits, the shoreline has retreated at an average rate of ~0.7 m/yr. The historical rate of shoreline retreat west of the Galveston Seawall averages −0.9 m/yr (Paine et al., 2021). At this historical rate of shoreline retreat, the island would have retreated ~2.0 km landward since 1.8 ka, which is clearly not the case given the current width of the island. Thus, the rate of landward retreat of Galveston Island has increased in historical time. This being the case, and considering net longshore drift is to the southwest, we should see an increase in the accumulation of sand within the San Luis Pass Tidal Delta, which is the depocenter for most of the sand that is eroded from the island. Wallace and Anderson (2013) tested this hypothesis by measuring the long-term accumulation of sand in the tidal delta.

The San Luis Pass Tidal Delta is composed of an extensive flood tidal delta and smaller ebb tidal delta (Figs. 23A, 23B). A detailed sediment facies analysis of the San Luis Pass Tidal Delta was conducted by Israel et al. (1987), who acquired 23 vibracores through the delta and mapped sedimentary facies. Wallace and Anderson (2013) re-occupied five of Israel's core sites at the fringe of the delta, where tidal facies are interbedded with bay mud (Fig. 23B). They acquired another five cores from the bay side of the west end of the island to examine the westward migration of the inlet and flood tidal delta. Radiocarbon ages of 3400 cal yr B.P. and 3200 cal yr B.P. were acquired from tidally influenced sediments in core SLPM4, which was collected ~5 km east of the current tidal inlet (Fig. 23A). From ca. 2.7 ka to ca. 2.1 ka, the inlet migrated west at least 2 km to the approximate modern location, yielding an average rate of lateral migration of 3–4 m/yr. Using historical charts, Wallace and Anderson (2013) estimated that the historical rate of inlet migration (over the past 200 years) was 7.5 m/yr, so the rate has roughly doubled in historical time. The current San Luis Pass Tidal Delta formed at ca. 2100 cal yr B.P., and its sand accumulation is directly related to the erosion of Galveston Island since that time. They estimated the sand flux from Galveston Island into San Luis Pass from ca. 2.1 ka to ca. 0.2 ka to be ~4700 m^3/yr. The long historical record of the migration and infilling of San Luis Pass Tidal Delta from navigation charts and sediment cores indicates that this flux has increased to ~10,000 m^3/yr over the past 200 years. These results support the interpretation of a recent increase in shoreline retreat rate based on the shoreface core data set and the argument for accelerated erosion of Galveston Island during historical time.

Follets Island. Follets Island is a relatively small barrier located west of Galveston Island, which is separated from it by San Luis Pass (Fig. 23). To the north of the island is Christmas Bay, a shallow (mostly less than 1.2 m water depth) bay with restricted connection to the gulf through San Luis Pass. The island lies above the eastern boundary of the Brazos River incised valley and is flanked on its eastern and western sides by Holocene Brazos River Channel belts. Sand supply to the island is minimal and is mainly through storm-driven recycling of sediments from the San Luis Tidal Delta (Morton et al., 1995). Sand supply from the west has been reduced by westward diversion of the Brazos River mouth from its pre-1929 location and the construction of jetties at the former river mouth at the west end of the island (Morton and Pieper, 1975). The limited sand supply to the island has resulted in a historical erosion rate that averages 1.9 m/yr, making it one of the fastest eroding shorelines of the Texas Coast (Morton, 1994; Paine et al., 2021). The modern rate of shoreline retreat at Follets Island is nearly equal to the rate of bay shoreline migration (Morton et al., 2004; Paine et al., 2012), which indicates that the barrier is in true rollover mode (Swift, 1968; Otvos, 2020).

A recent investigation of Follets Island by Odezulu et al. (2018) was based on 34 vibracores and 40 surface samples collected along four transects that extend from the upper shoreface across the island and into Christmas Bay (Fig. 25). These cores and surface samples were used to constrain the thickness of washover, barrier, and upper shoreface deposits and to estimate sediment fluxes toward the overall sand budget for the island over centennial to millennial timescales. CHIRP profiles from offshore show a thin upper shoreface unit resting on older strata (Carlin et al., 2015). Sediment cores from the upper shoreface and foreshore of the barrier sampled less than 1.5 m of sand resting on washover deposits, bay mud, and red clay (Fig. 25). The combined data indicate that washover deposits are a significant component of the island and support previous interpretations that Folletts Island is a transgressive barrier island (Bernard et al., 1970; Morton, 1994; Wallace et al., 2010).

A total of 21 radiocarbon ages were acquired from mostly articulated *Rangia cuneata* and *Ensis directus* to investigate the Holocene evolution of Folletts Island. These ages were acquired using the continuous flow gas bench accelerator mass spectrometer method at the Woods Hole Oceanographic Institution's NOSAMS facility, which yields less precise ages than the conventional AMS method. However, this approach allowed for more samples to be dated. In addition, ^{210}Pb analysis was conducted on two cores that sampled washover deposits to help constrain historical washover rates.

Radiocarbon ages from the Brazos River Delta and floodplain clays at the base of barrier sands and shoreface deposits are older than ca. 4100 cal yr B.P. (Fig. 25), which indicates that the barrier was seaward of its current location prior to this time. Bay muds began to accumulate between ca. 4200 cal yr B.P. and 3200 cal yr B.P. By ca. 3.0 ka, proximal washover deposits began accumulating on the bay side of the island, which suggests that the island was within 500–800 m of its current location by that time (Odezulu et al., 2018). The oldest shoreface deposits collected seaward of the island yielded ages in the range of

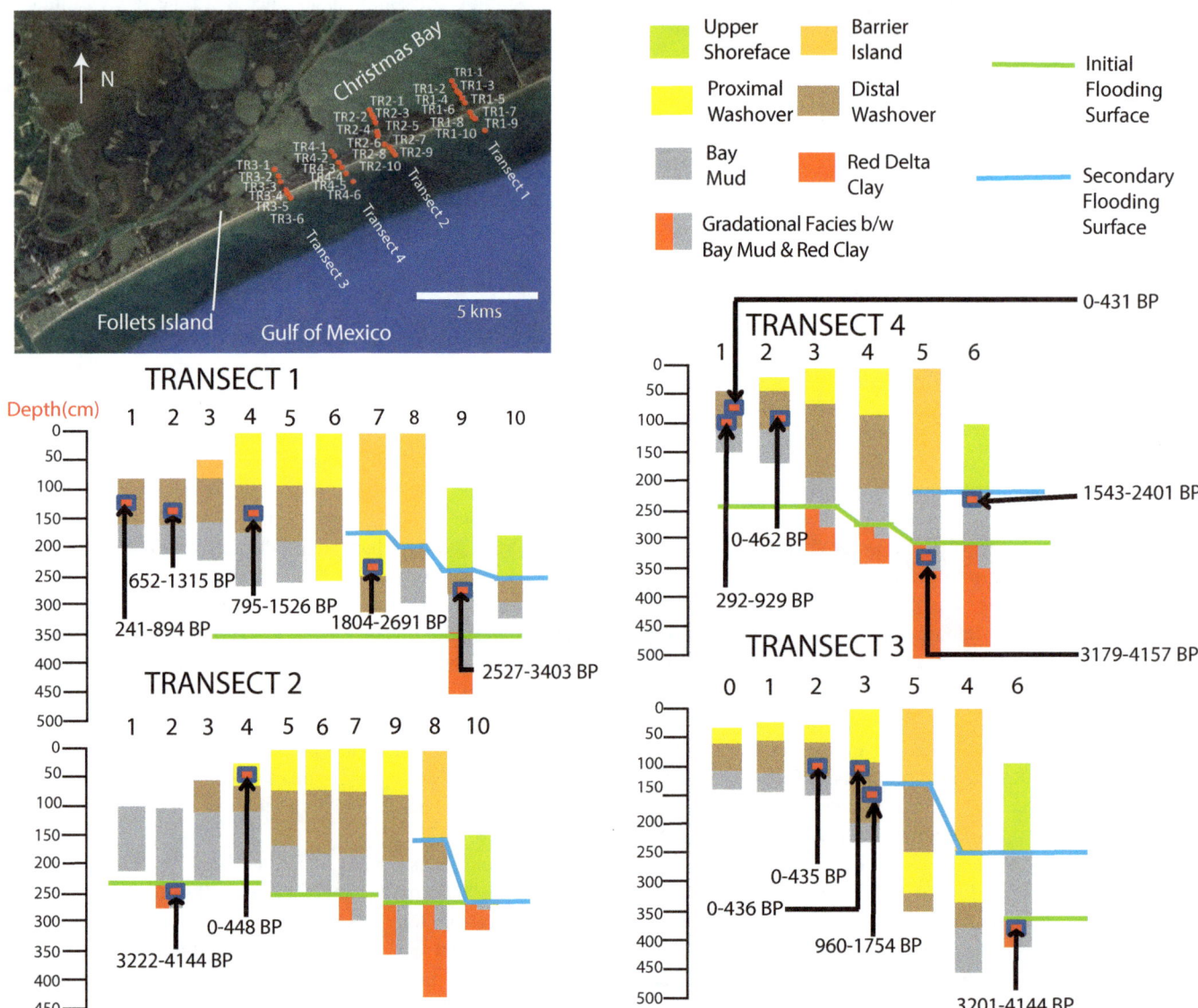

Figure 25. (Upper left) Locations of vibracores used by Odezulu et al. (2018) to study the Holocene evolution of Folletts Island are shown (base map from Google Earth). (Bottom panels) Sedimentary facies interpretations for four core transects across Folletts Island from the upper shoreface to the backshore bay show shoreface beach deposits above storm washover and bay deposits with the initial flooding surface (green) and secondary flooding surface (blue). Radiocarbon ages (in cal yr B.P.) are shown by the red boxes.

ca. 2400 cal yr B.P. to ca. 1500 cal yr B.P. These results are consistent with those from west Galveston Island and the San Luis Tidal Delta, which indicates that Follets Island had a history of relative stability before it began migrating landward. Odezulu et al. (2018) estimated the long-term rate of shoreline migration to be in the range of −0.17 m/yr to −0.27 m/yr during the past 3 ka, which is significantly slower than the historical rate of −1.9 m/yr (Paine et al., 2021).

To gain an independent estimate of the long-term versus short-term behavior of Follets Island, Odezulu et al. (2018) used historical charts, sediment cores and surface samples, and radiocarbon and ^{210}Pb ages to estimate the long-term and short-term overwash fluxes at Follets Island. The millennial timescale overwash flux was estimated to be ~2300 m^3/yr, as compared to the current flux of ~15,200 m^3/yr. Thus, the historical overwash flux is significantly greater than the long-term flux and accounts for at least half of the sand eroded from the barrier in historical time. Odezulu et al. (2018) estimate that at the current washover and shoreline erosion rates, Follets Island will likely cease to exist as a subaerial barrier in ~260 years.

In summary, the coastal barriers of the upper Texas Coast have had different styles and timing of evolution. This reflects differences in sand supply and the antecedent topography on which these barriers formed. Figure 26 illustrates this variability based on results from Bolivar Peninsula, Galveston Island, and Follets Island.

The Brazos Colorado Fluvial–Deltaic Systems

The combined Brazos and Colorado incised valleys span ~110 km of the Texas Coast and separate the broad, low-gradient east Texas continental shelf from the more narrow, higher gradient south Texas shelf (Fig. 9). Along with the Mississippi and Rio Grande rivers, they have been the dominant suppliers of fluvial sand to the Texas Coast, so understanding their response to climate change and sea-level rise during the Holocene is crucial to understanding coastal evolution in the region. Specifically, it is important to know how these rivers responded to different rates of sea-level rise and their history of avulsion, which controls the locations where sand is being delivered to the coast. Today, the Brazos and Colorado rivers nourish a broad deltaic headland, and their valleys are filled with fluvial sediments (Fig. 27). Detailed analysis of these sediments has yielded a rich stratigraphic record that reveals how these rivers responded to decreasing rates of sea-level rise and changing climate during the Holocene.

Offshore Brazos and Colorado Deltas. Seismic imaging of the Texas continental shelf revealed deltas associated with the Brazos, Colorado, and Rio Grande rivers that lie above the MIS 2 sequence boundary, so these deltas formed during the MIS 2–1 transgression (Snow, 1998; Abdulah et al., 2004; Banfield and Anderson, 2004). The Colorado and Rio Grande deltas, which were studied in the greatest detail, are separated by an embayment in the continental shelf that is filled with fine-grained sediments of the Texas Mud Blanket (TMB) (Fig. 28). Grids of high-resolution seismic data acquired from these deltas show that their more seaward portions are characterized by clinoforms whose thickness and variable orientations indicate lobe shifting. Thus, both were fluvial-dominated deltas during the early stages of their development (Snow, 1998; Abdulah et al., 2004; Banfield and Anderson, 2004). During the final stages of their development, both deltas changed from strongly lobate to more shore-parallel features (Figs. 28B–28C), which indicates a change from fluvial-dominated to wave-dominated deltas (Snow, 1998; Van Heijst et al., 2001).

Based on their water depths and locations on the shelf, Snow (1998) and Van Heijst et al. (2001) argued that both the Brazos and Colorado deltas began to form around 11.0 ka. In the case of the Brazos Delta, this age is supported by a single radiocarbon date (Abdulah et al., 2004). Global sea-level curves indicate moderate rise during the earliest Holocene, from ca. 11.2 ka to ca. 9.6 ka, followed by rapid rise, which is loosely referred to as meltwater pulse 1c (Fig. 4B). As previously discussed, this was about the time the shoreline shifted landward from Shepard Bank to Heald Bank and when initial flooding of the east Texas bays occurred. Thus, the change from fluvial-dominated to wave-dominated deltas is interpreted as having been caused by the 9.6 ka episode of rapid sea-level rise.

Onshore Brazos and Colorado Fluvial–Deltaic Systems: Brazos River Valley. Taha and Anderson (2008) used over 400 digital water well descriptions to map the Brazos River incised valley and characterize its fill. Their map of the valley shows that it increases from ~7 km wide to ~25 km wide as it crosses the low-gradient coastal plain (Fig. 29). Pleistocene (MIS 2) lowstand fluvial sands and gravels fill the more narrow, deeper portion of the valley. The upper, wider portion of the valley is filled mostly with floodplain silts and clays and five isolated channels. No estuarine deposits were sampled, and all five channels appear to extend to the modern coastline, so no marine flooding is recorded in the onshore valley. A stratigraphic section across the valley and located ~40 km inland was constructed by Abbott (2001) and provides chronostratigraphic constraints on the age of the valley fill (Fig. 30, section A–A′). This section reveals that the valley is filled mainly with Holocene deposits resting on Pleistocene sand and gravel that fills the deepest portion of the valley.

Recognizing that valley aggradation likely varies spatially and temporally along the valley axis, especially given changes in valley morphology, Taha and Anderson (2008) constructed an axial cross section that extends from the coast to ~80 km inland using data from six cores collected by Abbott (2001), Shell Oil Company (Bernard et al., 1970), and Sylvia and Galloway (2006), and two cores (BV-04-03 and BV-04-06) acquired as part of their study (Fig. 30, section B–B′). Existing radiocarbon dates from several cored sites (Abbott, 2001; Sylvia and Galloway, 2006) augmented by dates from Taha and Anderson (2008) provide age constraints for the Brazos valley fill. The combined results show that the valley fill spans the late Pleistocene through middle Holocene with two episodes of rapid aggradation at ca. 11.5 ka and 9.5 ka, coinciding with periods of rapid sea-level rise. They cautioned that other factors, in particular changes in offshore

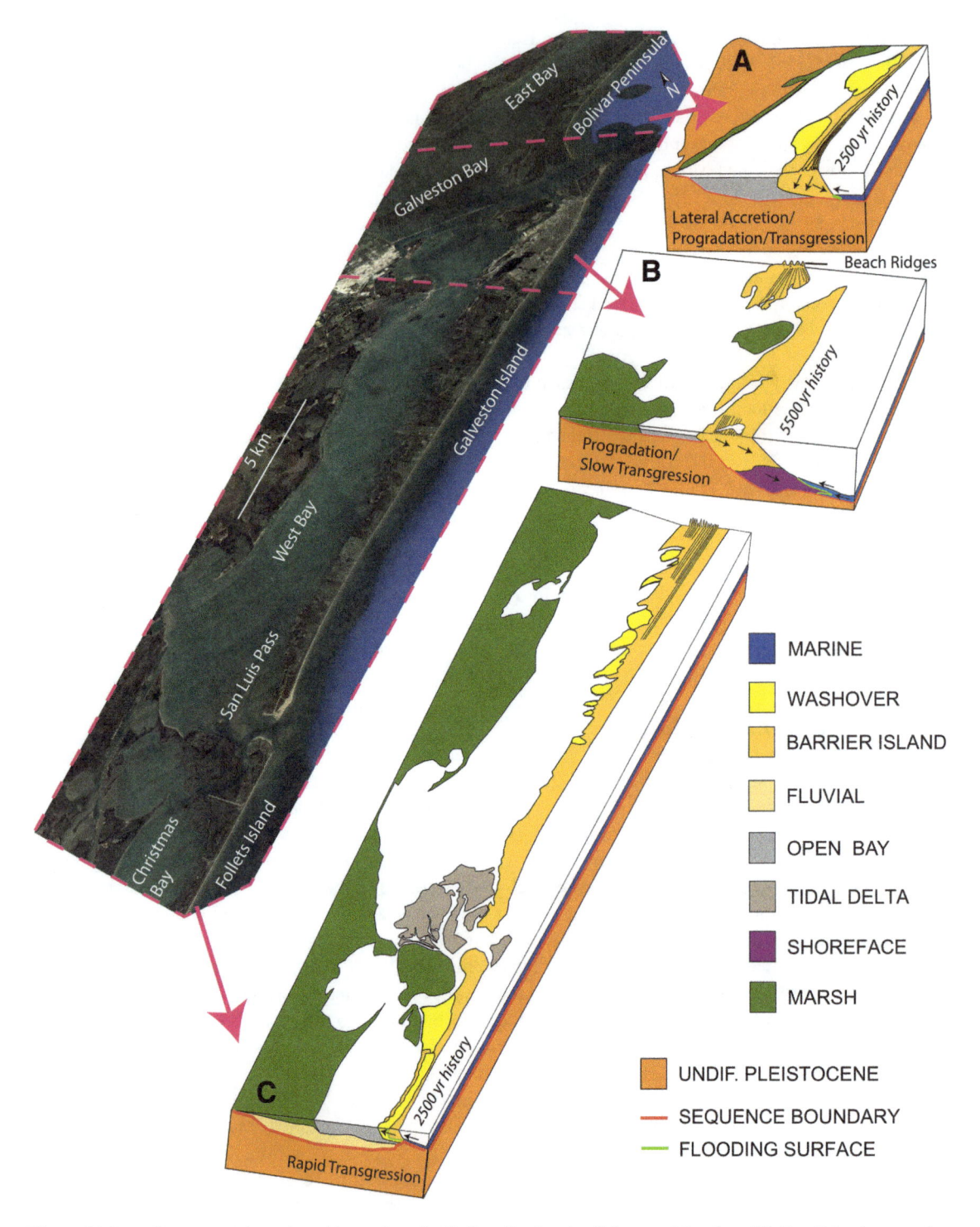

Figure 26. Box diagrams and stratigraphic sections for Bolivar Peninsula, Galveston Island, and Follets Island are used to illustrate variable styles of barrier evolution along the east Texas Coast (base map from Google Earth). (A) Bolivar Peninsula is a regressive barrier whose growth was interrupted by a brief transgressive event that removed beach ridges older than ca. 0.8 ka. The peninsula experienced a later phase of progradation at ca. 0.65 ka. (B) Galveston Island is mainly a regressive barrier, but its younger beach ridges have been eroded. This later transgressive phase of barrier evolution is recorded in offshore cores by the onlap of marine mud onto shoreface deposits. (C) Follets Island rests on back-barrier bay mud and washover deposits, which is indicative of a transgressive history. Note that the scales are different, and topography is not shown because cores were collected at or near sea level.

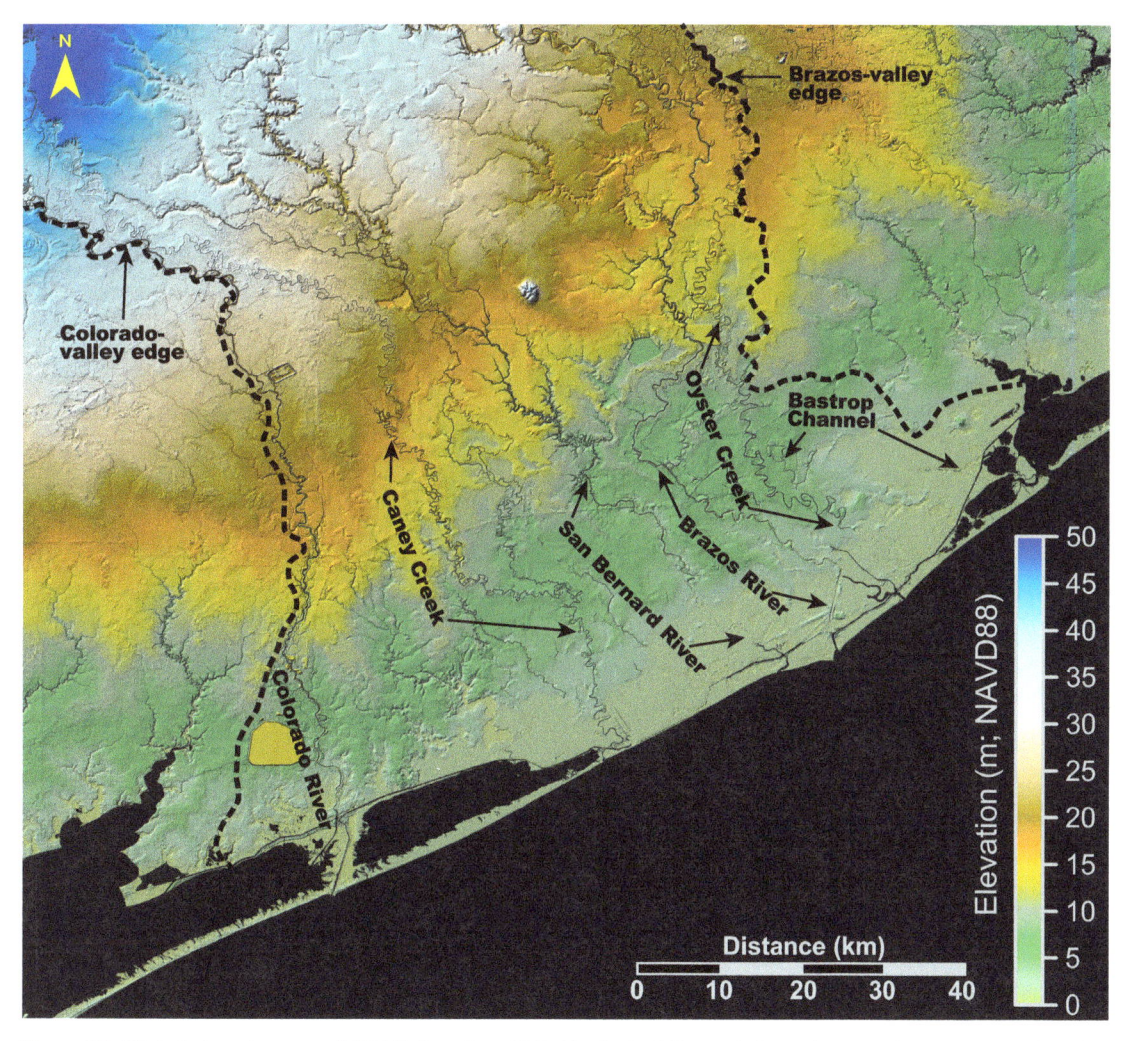

Figure 27. Digital elevation map (LIDAR data from U.S. Geological Survey) shows latest Holocene and modern fluvial channels within the Brazos-Colorado valley (heavy dotted lines show margins).

gradient, also influenced delta construction and retreat. Fluvial aggradation rates from ca. 9.5 ka to ca. 6.0 ka in the lower portion of the valley averaged ~2.0 m/ka, filling most accommodation within the valley at about the rate of sea-level rise. Since ca. 6.0 ka, aggradation has shifted northward to the upper, narrower portions of the valley, where accommodation is significantly less than in the lower, wider portions of the valley. The slow rate of aggradation in the lower portion of the valley since ca. 6.0 ka implies sediment bypass of the valley and higher discharge to the Gulf during the late Holocene.

Another factor to consider with regard to fluvial sediment delivery to the coast during the Holocene is the role of river avulsion in controlling the location of the river mouth at the coast. Given existing age constraints, Taha and Anderson (2008) concluded that channel avulsion for the Brazos River occurred approximately every 2.4 ka. As aggradation of the valley progressed and the narrower portions of the valley were filled, the spacing between channels increased. The lowstand channel and three early Holocene channels are confined to the deeper, narrower central portion of the valley (Fig. 29), and their sediment loads were focused on the location of the offshore early Holocene delta. The younger Bastrop and Oyster Creek channels are located near the eastern margin of the incised valley and are still prominent landscape features (Fig. 27). Thus, the location of sediment delivery to the coast from the Brazos River during the early Holocene shifted nearly 20 km from the center of the valley to the eastern margin of the valley.

A bathymetric feature at −12 m to −16 m, known as the Big Slough Bathymetric High, occurs offshore the Bastrop Channel. A number of sediment cores that sampled this feature recovered sediments that resemble those collected offshore of the modern Brazos Delta (Siringan and Anderson, 1994; Rodriguez et al., 2004). Sediment cores collected from San Luis Pass, Follets Island, West Bay, and Christmas Bay also sampled red clay that is interpreted as Brazos floodplain and delta plain deposits (Bernard et al., 1970; Odezulu et al., 2018). These deposits are in

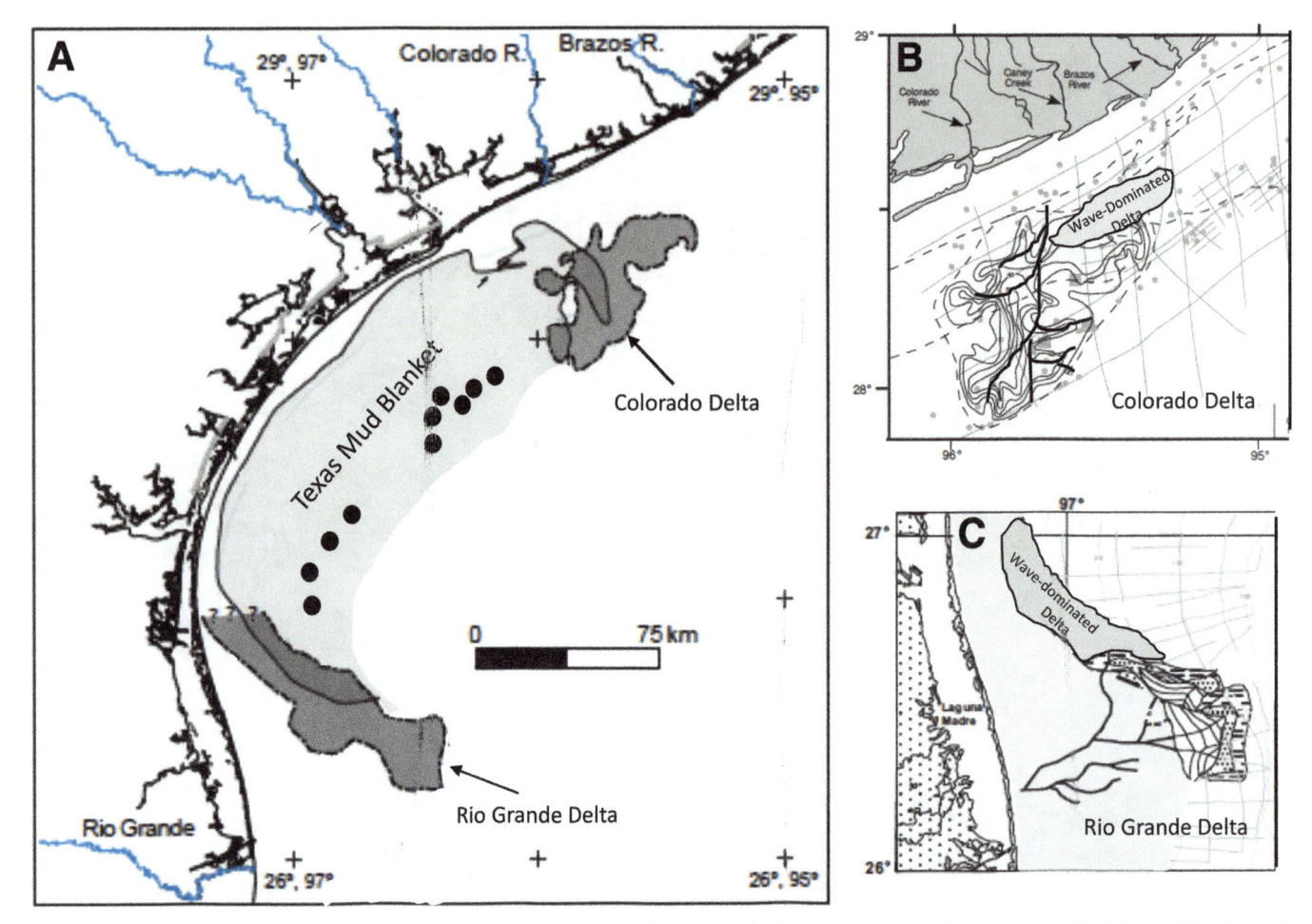

Figure 28. (A) Paleogeographic map shows offshore Colorado Delta, Rio Grande Delta, and Texas Mud Blanket (modified from Odezulu et al., 2020). Black dots are approximate locations of coralgal reefs. (B) Isopach map of Colorado Delta shows lobate shape of more offshore delta and more landward shore-parallel, wave-dominated delta (modified from Snow, 1998). (C) Paleogeographic map of Rio Grande Delta shows more offshore lobate delta and more onshore wave-dominated delta (modified from Banfield and Anderson, 2004).

sharp contact with overlying bay, washover, and tidal deposits. Radiocarbon ages of a peat sample directly beneath the red clay and from bay muds just above this contact indicate that the river occupied Bastrop Channel between ca. 7.5 ka and ca. 4.0 ka (Siringan and Anderson, 1994; Odezulu et al., 2018). Avulsion to the Oyster Creek Channel, as indicated by the oldest ages from floodplain and delta plain deposits, occurred ca. 4.0 ka, and the channel was active until ca. 1.5 ka (Aten, 1983; Bernard et al., 1970). The Oyster Creek Channel was imaged by a ground-penetrating radar profile collected ~0.5 km inland of the modern shoreline, but sediment cores and seismic lines collected directly offshore of this location show no bathymetric or sedimentary remnants of a delta. Rather, seismic profiles and cores reveal marine mud resting on a Pleistocene surface. Thus, unlike the Bastrop Delta, the Oyster Creek Delta has been completely eroded. Following the Oyster Creek avulsion at ca. 1.5 ka, a new delta formed ~3 km west of Oyster Creek at Surfside Beach, Texas (Fig. 31). This delta was a prominent feature when the first maps and navigation charts for the area were constructed, with an offshore lobe discernable to a water depth of −12 m. It has subsequently been eroded.

Rodriguez et al. (2000) acquired 96 vibracores from both the onshore and offshore portions of the modern Brazos Delta and used these cores to characterize sedimentary facies and the stratigraphic architecture of the delta. One of their key observations was that the post-1960 onshore delta is not dominated by sand, but rather composed of a succession of seaward prograding headlands composed of thin sand ridges separated by silt and clay. They also found that there is a significant offshore component of the delta that is composed dominantly of silt and clay. Thus, the bulk of the delta is composed of fine-grained sediment and reflects the fact that the Brazos is a suspended load-dominated river. The sand component is dominated by fine to very fine sand, the exception being coarser channel and mouth bar sands. USGS stream gauge records for the Brazos River at Richmond, Texas, extend back to 1927 and show that there have been 22 flood-related sediment discharge events when velocities were sufficient to transport fine to very fine sand in suspension. Thus,

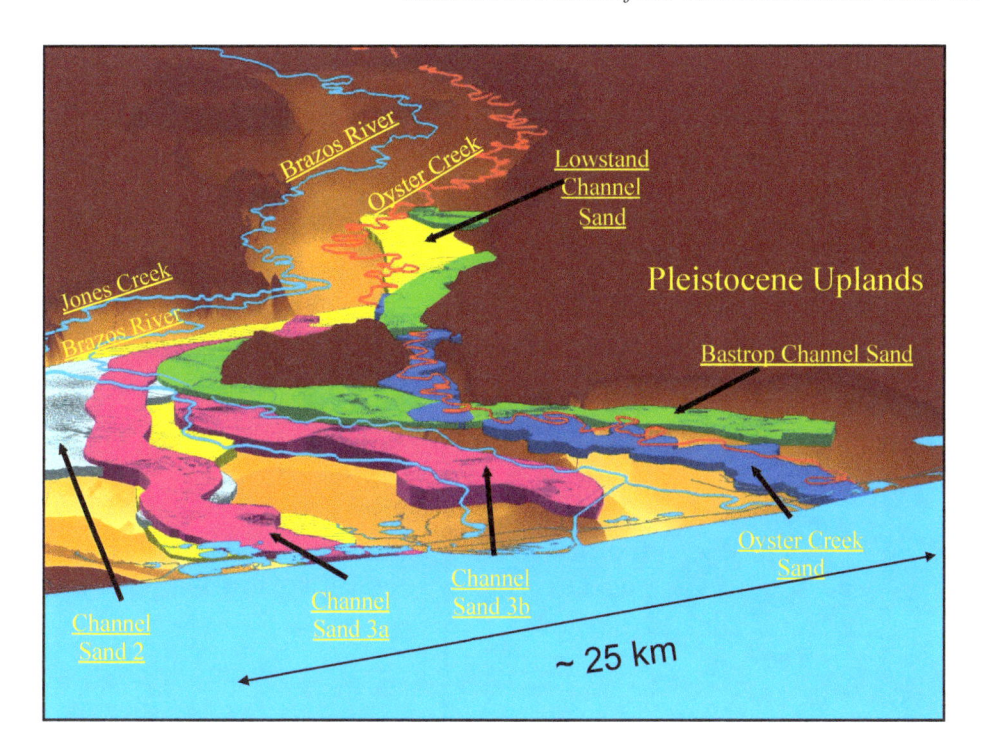

Figure 29. Oblique, 3-D view of the Brazos incised valley shows individual Holocene channels. Map was constructed by Taha and Anderson (2008) using over 400 water well descriptions (modified from Anderson et al., 2016).

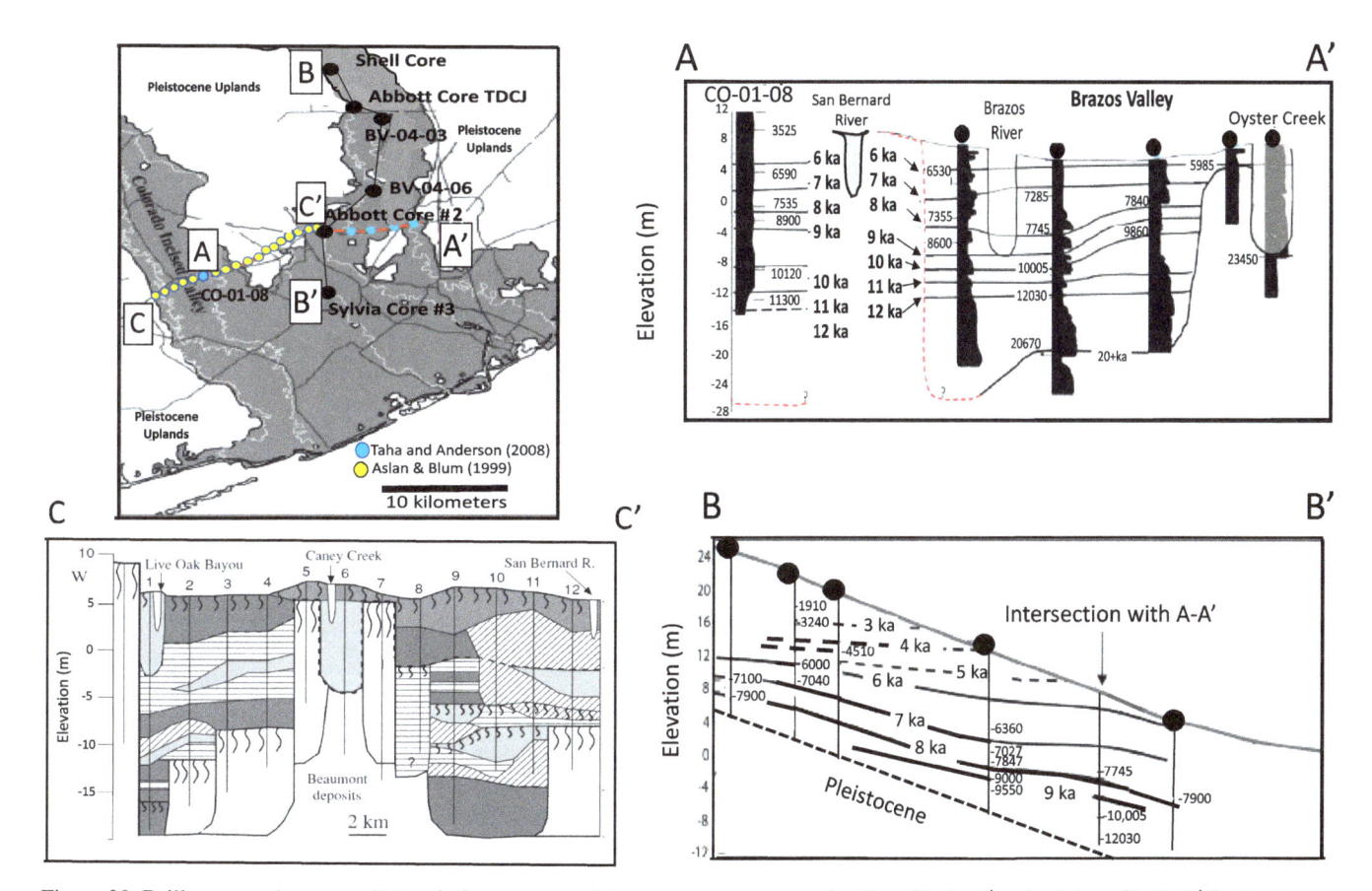

Figure 30. Drill cores and water well descriptions were used to construct a cross-sectional profile A–A′ and axial profile B–B′ for the Brazos valley (modified from Anderson et al., 2016, and Taha and Anderson, 2008, respectively) with isochrons used to study valley aggradation history. Section C–C′ (yellow dots) of the Colorado valley was constructed by Aslan and Blum (1999). As at the Brazos valley, the number of channels is similar, and they are packaged within mostly fine-grained floodplain deposits. Radiocarbon ages from a single core (CO-1-08) indicate a similar aggradation history for the two valleys.

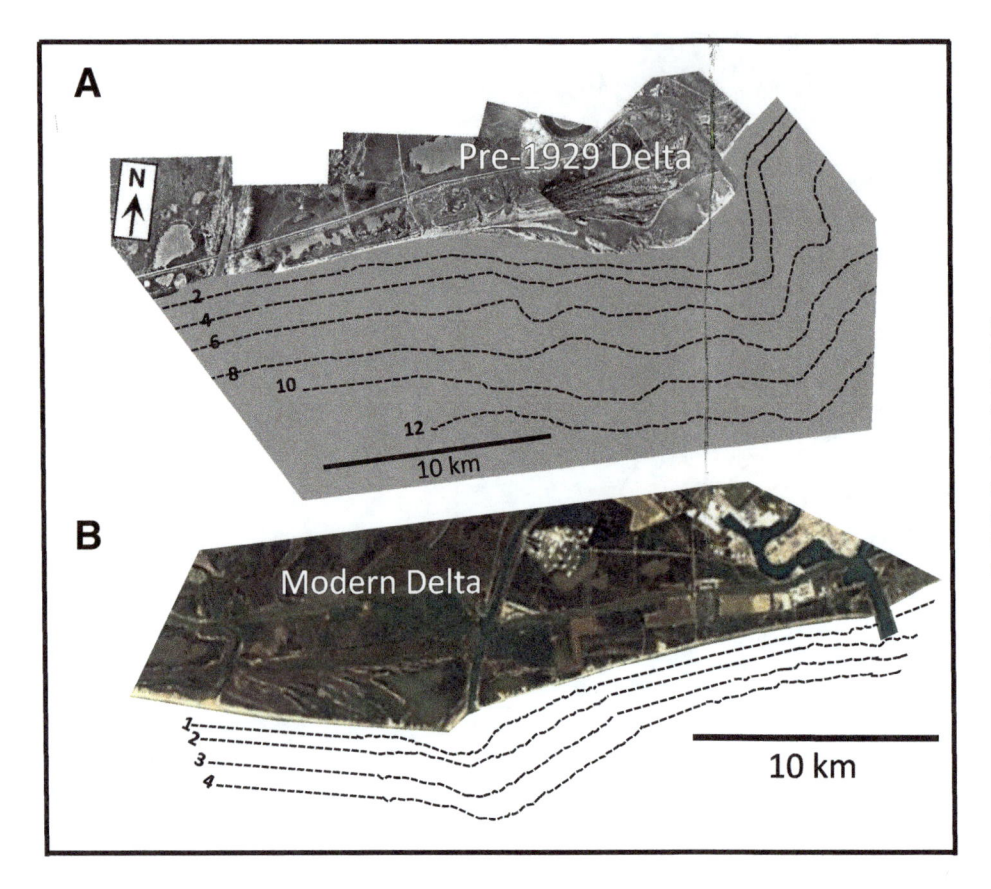

Figure 31. Aerial photographs and bathymetric maps of the (A) pre-1929 and (B) modern Brazos deltas show changes that occurred after the river was diverted to the west. The contour interval for the pre-1929 delta is in feet, and the contour interval for the modern delta is in meters. Note that the pre-1929 delta has mostly eroded. Both figures are modified from Rodriguez et al. (2000).

a significant component of the sand is transported to the coast in suspension during high-discharge events, which indicates that sand supply to the coast is regulated by climate (Rodriguez et al., 2000).

The Colorado River Valley. The drainage basins of the Brazos and Colorado rivers lie adjacent to one another, are similar in size (118,000 km^2 and 110,000 km^2, respectively), and experience similar climate regimes (Fig. 3). The geology of the two drainage basins differs within the central portion of the state, where the Brazos River crosses mostly terrigenous sedimentary strata and the Colorado River crosses more carbonate strata and crystalline rocks of the Llano Uplift. This is manifest as somewhat coarser and more mineralogically immature sands and gravels in the Colorado River.

Blum (1993) and Aslan and Blum (1999) compiled a stratigraphic section across the Colorado River valley using drill sites and sand pit exposures (Fig. 30C). Their results show a similar stratigraphic character to that of the Brazos River valley, with a similar number of isolated channels within mostly fine-grained floodplain deposits. We acquired a single drill core (CO-01-08; Fig. 30A) near the western side of the valley aimed at acquiring radiocarbon ages to compare the aggradation histories of the Brazos and Colorado valleys. The results from six radiocarbon ages from this drill site indicate that the rate of aggradation within the lower Colorado valley is similar to that of the Brazos valley and that there is no evidence for marine incursion at the location of the drill site. The lower portions of both valleys appear to have experienced slower rates of aggradation during the late Holocene as the rate of sea-level rise decreased and fluvial aggradation shifted up-valley (Fig. 30, B–B′). The southern portions of both valleys are filled with sediment. This implies that, prior to human alterations to the channel, more of the sediment discharge of both rivers was bypassing the lower alluvial plain and making its way to the Gulf of Mexico. This is consistent with results from a study of the offshore Texas Mud Blanket by Weight et al. (2011), who observed a significant expansion of the mud blanket after ca. 4.0 ka. Although most of that sediment came from the Mississippi River, mineralogical analyses indicate that the Brazos and Colorado rivers also contributed to the mud blanket's late Holocene expansion.

Given a lack of age constraints for the older channels within the lower Colorado valley, it is not possible to reconstruct its channel avulsion history, but the same number of channels occurs in both valleys, which indicates a similar Holocene avulsion frequency (approximately every 2.4 ka) for the two rivers. It is notable that the most recent avulsion of the Colorado River took place ~1000 years ago (McGowen and Brewton, 1975) and resulted in the river mouth shifting ~40 km to the west of its former location at the mouth of Caney Creek (Fig. 27). There is no onshore geomorphic or bathymetric expression of a delta at the Caney Creek location, so similar to the Oyster Creek and the pre-1929 Brazos deltas, the Caney Creek Delta was likely eroded. Complete erosion of these late Holocene deltas underscores the efficiency of transgressive erosion during slow sea-level rise.

In summary, sediment supply to the Gulf of Mexico from the Brazos and Colorado rivers was large enough during the early Holocene for these rivers to construct fluvial-dominated deltas on the continental shelf, despite the relatively high rate of sea-level rise. These deltas were abandoned at ca. 9.6 ka, probably in response to the same punctuated sea-level event that led to the initial flooding of incised river valleys in the region. This was followed by a brief episode of wave-dominated delta formation as fluvial sedimentation shifted to the onshore Brazos and Colorado incised valleys. Since then, aggradation in the valley has increased, and the locus of sedimentation has shifted to the north. There is no evidence for either valley having been flooded during the Holocene or evidence for a significant change in sediment supply from these rivers to the coast, but avulsion has altered the locations of the river mouths, and more so for the Colorado River than the Brazos River.

The Central Texas Coast

The area of the central Texas Coast extends from Matagorda Peninsula to Baffin Bay (Fig. 1). The region is characterized by a subhumid climate and rivers with relatively small drainage basins and low water and sediment discharge rates (Fig. 3). However, the central Texas shelf embayment is the depocenter for the Texas Mud Blanket, which is sourced mainly from the Mississippi River and is the thickest accumulation of Holocene marine mud in the LaTex region. As we will see, the accumulation of this mud blanket has had a major influence on coastal barrier evolution in the area.

The Texas Mud Blanket. During the early MIS 2–1 transgression, fringing coralgal reefs occupied a 120-km-long stretch of the outer continental shelf (Fig. 28). Detailed swath bathymetric mapping of these reefs led to the discovery of six distinct terraces that were interpreted as having formed during times of punctuated sea-level rise (Khanna et al., 2017). Based on their depth distributions, these punctuated sea-level events occurred between 14.5 ka and 12.5 ka. During this time of reef growth, sedimentation rates on the central Texas shelf were low, which suggests that sediment supply was accordingly low (Weight et al., 2011). This was followed by a regime shift in sedimentation as evidenced by deposition of the Texas Mud Blanket (TMB), a massive (~300 km^3) accumulation of mud that occupies the embayment in the continental shelf between the Rio Grande and Colorado paleo-deltas (Shideler, 1978; Weight et al., 2011; Fig. 28).

Weight et al. (2011) conducted a detailed analysis of the TMB using a grid of high-resolution seismic data and several oil company platform borings. Their analysis included a robust radiocarbon age model that allowed for a detailed history of its growth and expansion. During the initial MIS 2 transgression, mud accumulation on the shelf was minimal, with an average accumulation rate of 0.2 mm/yr, which indicates low turbidity that was conducive to reef development. By the early Holocene, the rate of mud accumulation increased by an order of magnitude to an average rate of 2.0 mm/yr, which was followed by a period of reduced accumulation (0.3 mm/yr) between ca. 7.5 and ca. 4.0 ka. This latter event coincided with the mid-Holocene Climate Optimum, which was a period of warmer and dryer conditions in central Texas (Toomey et al., 1993; Nordt et al., 1994, 2002; Wong et al., 2015) and highlights the importance of climate in controlling sediment delivery and dispersal to the region.

The late Holocene was a period of rapid growth of the TMB, when its volume increased ~172 km^3 (Weight et al., 2011). The impact of this rapid mud accumulation was the continued burial of reefs, which reduced them to pinnacle reefs (Fig. 32), and burial of the ancestral Rio Grande and Colorado deltas. Mineralogical data indicate that the main sediment source of the TMB was the Mississippi River, although local sources (the Brazos, Colorado, and Rio Grande rivers) also contributed. Weight et al. (2011) concluded that the delivery of this sediment to the central Texas shelf from the Mississippi, Brazos, and Colorado rivers was controlled by the southeastern wind-dominated alongshore transport, coupled with the Louisiana–Texas Coastal Current (Sionneau et al., 2008; Fig. 5). A recent study by Carlin et al.

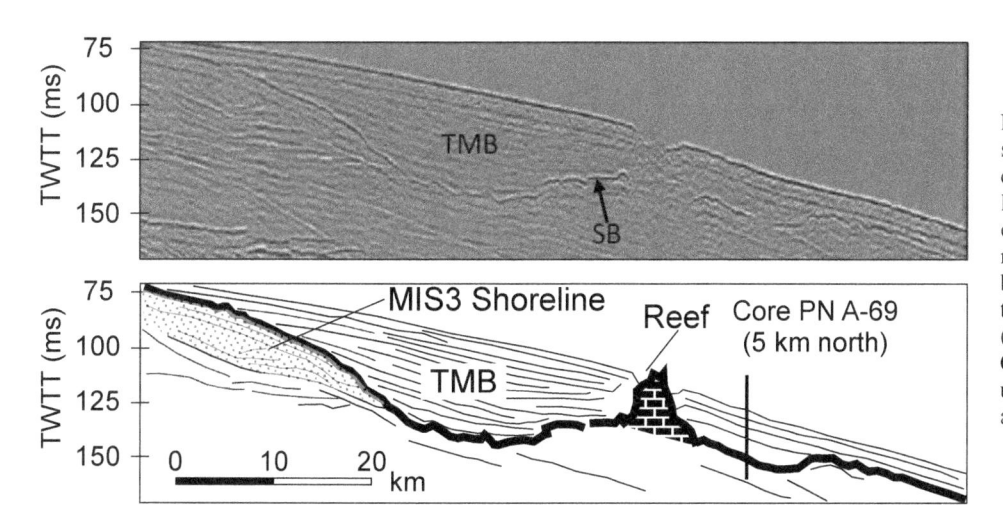

Figure 32. Uninterpreted and interpreted seismic section from the central Texas continental shelf shows the acoustically laminated Texas Mud Blanket (TMB) overlying an irregular surface, the marine isotope stage (MIS) 2 sequence boundary (SB), and one of several reefs that were nearly buried by the TMB (modified from Weight et al., 2011). Core PN A-69 is one of several cores used by Weight et al. (2011) to measure accumulation of the TMB.

(2021) focused on the more recent development of the TMB and showed that it may contain a record of recent flood events from larger Texas rivers.

Holocene Evolution of Bays. The central Texas area includes several bays (Matagorda Bay, San Antonio Bay, Copano Bay–Aransas Bay, and Corpus Christi Bay; Fig. 1). Early studies of San Antonio Bay (Shepard and Moore, 1955; Parker, 1959) and Aransas Bay (Behrens, 1963) were based on drill cores that penetrated these valleys and revealed a stratigraphic section indicative of greater marine influence upward in the section. More recent studies that relied on high-resolution seismic data and drill cores along with radiocarbon ages have since provided more detailed records of the evolution of Copano Bay (Troiani et al., 2011), Matagorda Bay (Maddox et al., 2008), and Corpus Christi Bay (Simms et al., 2008). The results from these studies show that, despite the relatively small drainage basin areas and low discharges of central Texas rivers, they occupy relatively large and deep incised valleys with expanded stratigraphic records of Holocene sea-level and climate change.

Matagorda Bay. The Matagorda Bay Complex consists of Matagorda Bay and five auxiliary bays: Lavaca, Keller, Caranchua, Cox, and Tres Palacios bays (Fig. 33). Lavaca Bay has an average depth of 2 m and Matagorda Bay an average depth of 4 m. The bay complex experiences a mean tidal range of 43 cm but is strongly influenced by wind tides with amplitudes of up to 1.0 m. The Lavaca River is the main sediment source for the bay system, and the Navidad River is the second largest source. These are coastal plain rivers that have a combined drainage basin area of just over 6000 km^2, with the headwaters of both rivers occurring within 120 km of the coast. The Lavaca River is the only one of these rivers to have constructed a bayhead delta in the upper portion of Lavaca Bay in modern time.

A relatively dense grid of sparker seismic data acquired by the U.S. Geological Survey (Byrne, 1975) guided our boomer seismic survey, and the combined data were used to map the Pleistocene surface within the valley (Fig. 33). The incised valleys of the different rivers and streams converge seaward into one large valley with a deep (MIS 2) incision near the center of Matagorda Bay that extends under the western end of Matagorda Peninsula. The smaller valleys are now occupied by the auxiliary bays.

Similar to bays of the east Texas Coast, seismic records from the Matagorda and Lavaca bays show distinct facies bounded by sharp flooding surfaces in which more seaward facies are superimposed on more landward facies (Fig. 34). Five drill cores were acquired along the central axis of the incised valley, spaced at approximately equal distances from the northern portion of Lavaca Bay to ~1.5 km landward of Matagorda Peninsula. These cores were used to correlate seismic facies and lithofacies and obtain radiocarbon ages to constrain the history of bay evolution (Maddox et al., 2008). The results were then correlated to the stratigraphic section of Wilkinson and Basse (1978), which is based on a shore-parallel series of borings collected near the bay side of Matagorda Peninsula. A total of 22 radiocarbon dates was used to constrain the ages of flooding surfaces within the bay.

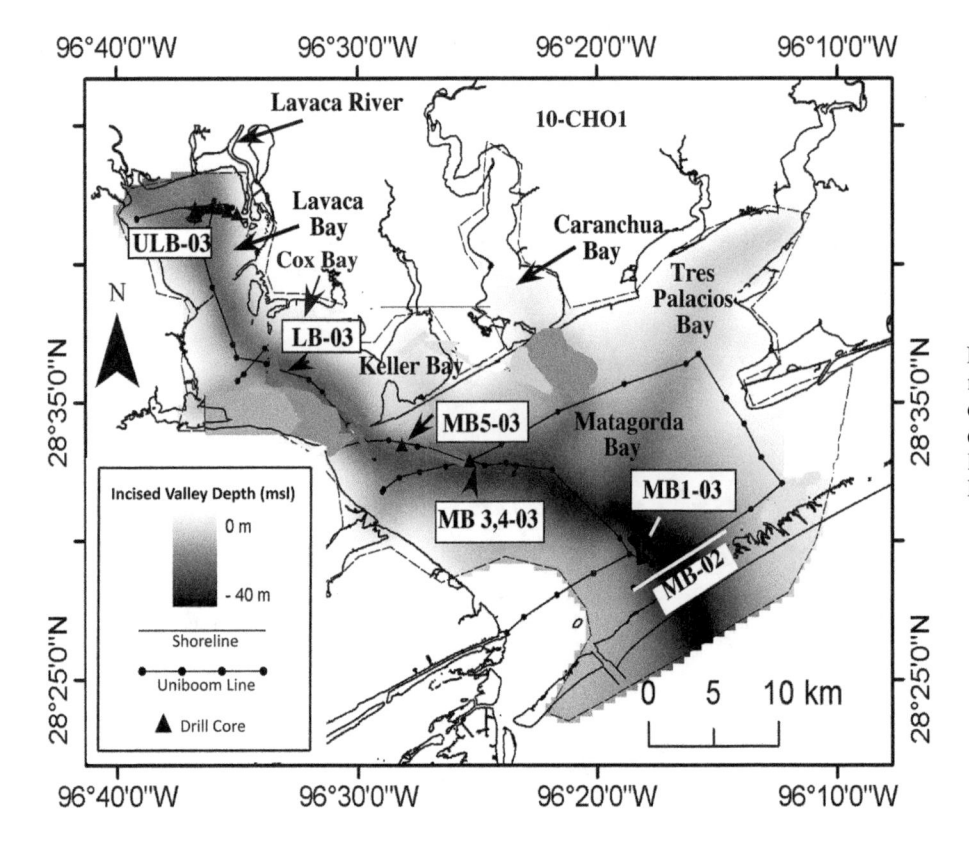

Figure 33. Shaded relief and geographic map of the Lavaca incised valley shows locations of Rice University seismic lines and drill cores used to study the Holocene evolution of Matagorda Bay (modified from Maddox et al., 2008).

A bayhead delta and marshes occupied much of the Lavaca valley following initial flooding at ca. 9.6 ka, followed by slow bayhead delta progradation (Fig. 34). At ca. 8.2 ka, a flooding event resulted in a northern shift in the Lavaca bayhead delta of ~10 km. Following the ca. 8.2 ka flooding event, the bay changed little until ca. 7.5 ka, when another flooding event occurred, culminating in an open bay setting in the southern portion of the valley and the formation of prominent spits that mark the location of an inlet near the center of the valley (Fig. 35). This inlet was ~1.6 km landward of the current shoreline, and based on clinoform thickness, it was up to 10 m deep. The age of these spits was constrained using drill core MB1-03. This core yielded a radiocarbon age of 7370 cal yr B.P. near the base of the accretionary unit and an age of 6700 cal yr B.P. from the overlying bay mud.

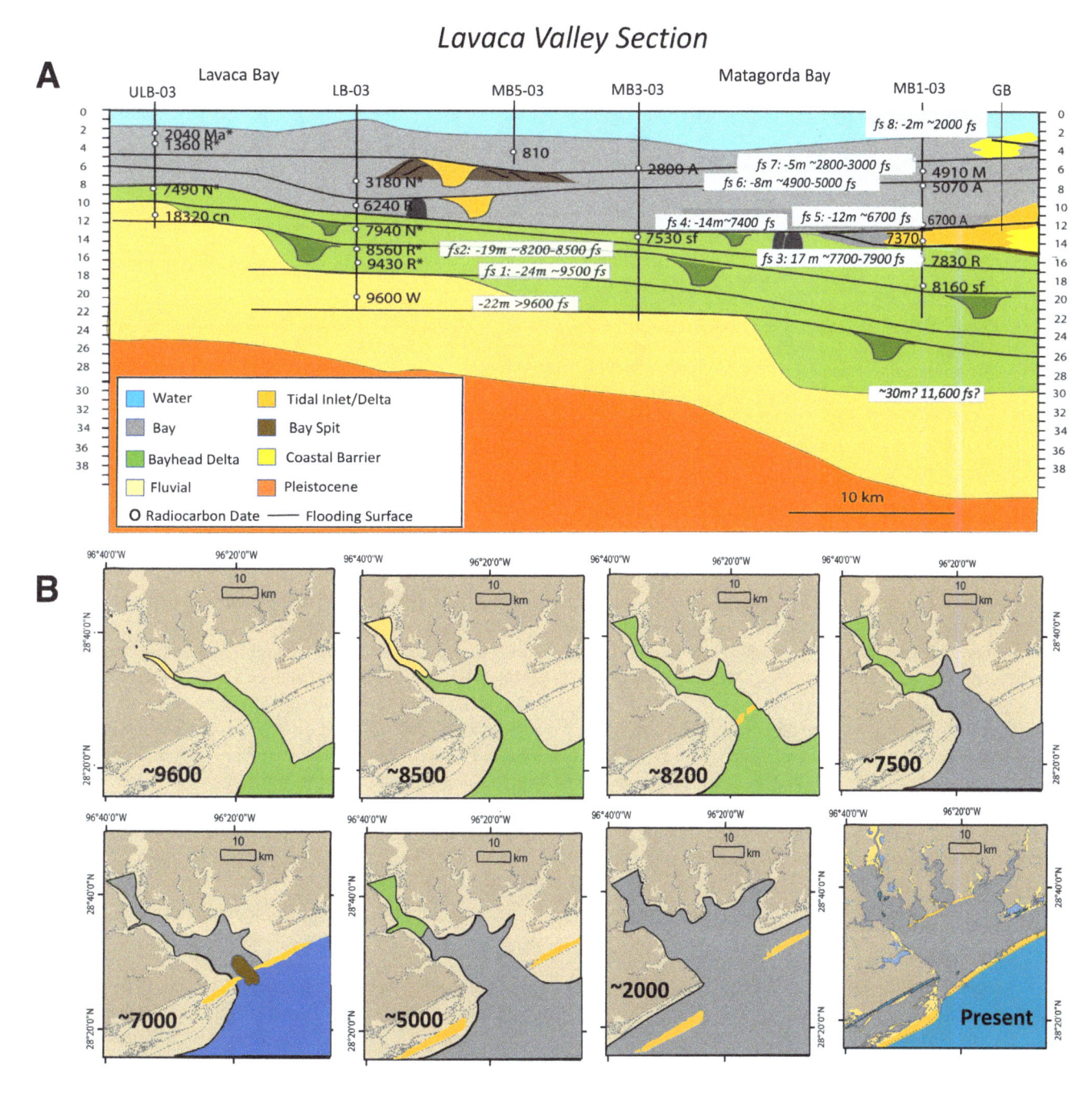

Figure 34. (A) Matagorda Bay north–south stratigraphic section compiled from seismic data and drill cores collected along the axis of the Lavaca incised valley (see Fig. 33 for locations of drill cores). Also shown are flooding surfaces (fs) that mark episodes of landward-shifting bay facies, with radiocarbon ages used to constrain the ages of these surfaces. (B) Paleogeographic maps depict changes in bay evolution based on data shown in the above cross section (modified from Maddox et al., 2008). Dates are given in cal yr B.P.

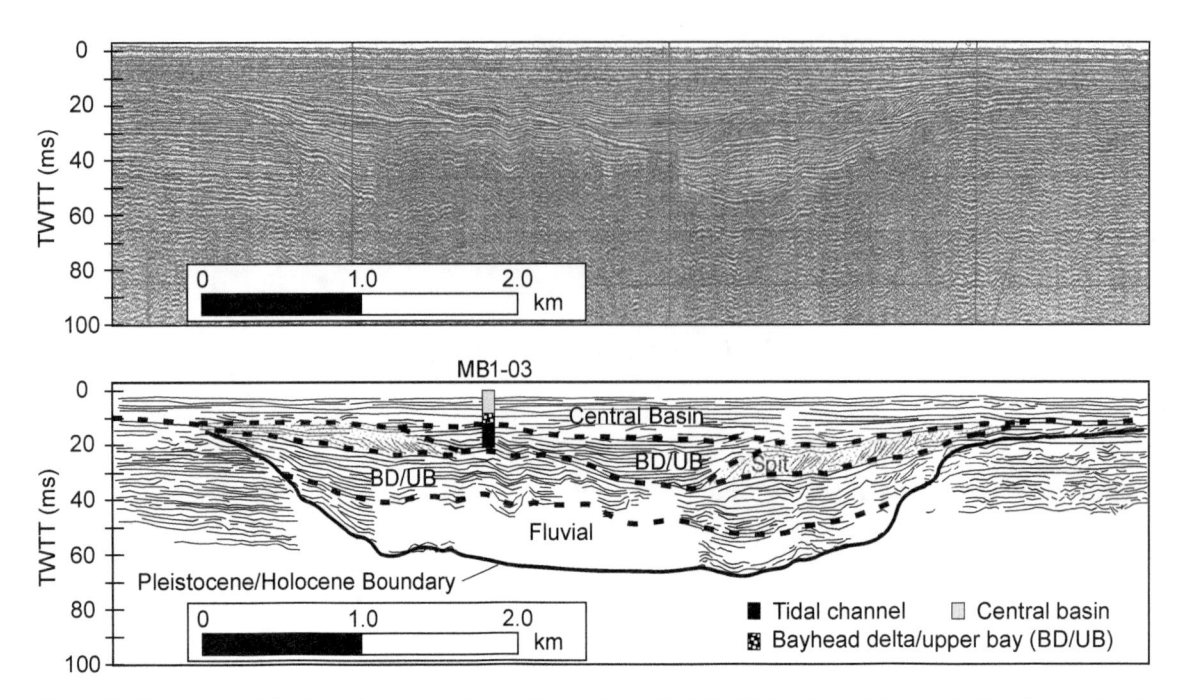

Figure 35. Uninterpreted (top) and interpreted (bottom) seismic profile MB-02 from lower Matagorda Bay shows seismic units and major flooding surfaces. Note accretionary clinoforms that are interpreted as spits that formed when the shoreline was near this location (modified from Maddox et al., 2008).

By ca. 5.0 ka, sea level had reached the shallow flanks of the valley, and the bay began a phase of significant widening and progradation of the bayhead delta and by ca. 2.0 ka, the bay had reached its current shape (Fig. 34). Tidal inlet and delta deposits were sampled by Wilkinson and Basse (1978) at the western end of the peninsula in the vicinity of Pass Cavallo and record a final stage of tidal inlet formation at this location after ca. 2 ka (Fig. 35).

Copano Bay. Unlike other bays of the Texas Coast, Copano Bay is aligned roughly parallel to shore. It occupies only the northern portion of the onshore Aransas–Mission River incised valley with Aransas Bay occupying the lower half of the valley. Troiani et al. (2011) used high-resolution seismic data and six sediment cores to investigate the role of climate and sea-level changes on the evolution of Copano Bay. Their cores contain a record that spans most of the Holocene since initial flooding of the bay at ca. 9.6 ka, which was followed by punctuated episodes of environmental change. The most dramatic environmental change occurred at 8.2 ka when the bayhead delta shifted landward 7.5 km. Using grain-size data to develop a wind proxy, they recognized a period of relatively strong winds between 5.2 ka and 4.1 ka, followed by calmer winds during the late Holocene, which is consistent with other paleoclimate records for the region linking higher winds to drier conditions.

Corpus Christi Bay. Corpus Christi Bay and Nueces Bay compose a single elongate estuary that occupies the lower reaches of the Nueces River incised valley. It is a microtidal estuary with average depths of 3–4 m in Corpus Christi Bay and 1 m in Nueces Bay. The Nueces River is the principal source of freshwater and sediments to the bay system. Its drainage basin is characterized by a semiarid climate, and evaporation exceeds precipitation (White et al., 1983). The bay is connected to the Gulf by Aransas Pass, which is located at the eastern end of the bay (Fig. 36A), and historically by Fish Pass and Corpus Christi Pass.

Simms et al. (2008) acquired a grid of high-resolution seismic data and combined these data with existing lines acquired by the USGS in 1978 and 1996, coupled with descriptions of over 90 platform borings from the U.S. Army Corp of Engineers and Texas Department of Transportation, to map the Nueces incised valley (Fig. 36B). Their map shows a narrow valley that exceeds 40 m deep at the coast with broad terraces on both sides of the valley. They also acquired 23 pneumatic hammer cores and 12 rotary drill cores to examine lithofacies and stratigraphy of the bay and 53 radiocarbon ages to examine the valley's flooding history. As in all other bays, the stratigraphic architecture of the Nueces valley is characterized by landward stepping bayhead delta, middle bay, and tidal delta facies separated by distinct flooding surfaces (Fig. 37).

Initial flooding of the Nueces valley occurred prior to ca. 9.7 ka. The most prominent flooding events occurred at ca. 8.2 ka, ca. 4.8 ka, and ca. 2.6 ka (Simms et al., 2008; Ferguson et al., 2018a). During the ca. 8.2 ka flooding event, the area of the middle bay more than doubled, and the bayhead delta shifted landward ~15 km. Ferguson et al. (2018a) observed a five-fold increase in dinoflagellate cysts associated with this flooding event, which indicates an abrupt increase in bay salinity that resulted in open marine conditions and destroyed oyster reefs within the bay (Simms et al., 2008; Goff et al., 2016). They argued that this magnitude of change could only have occurred

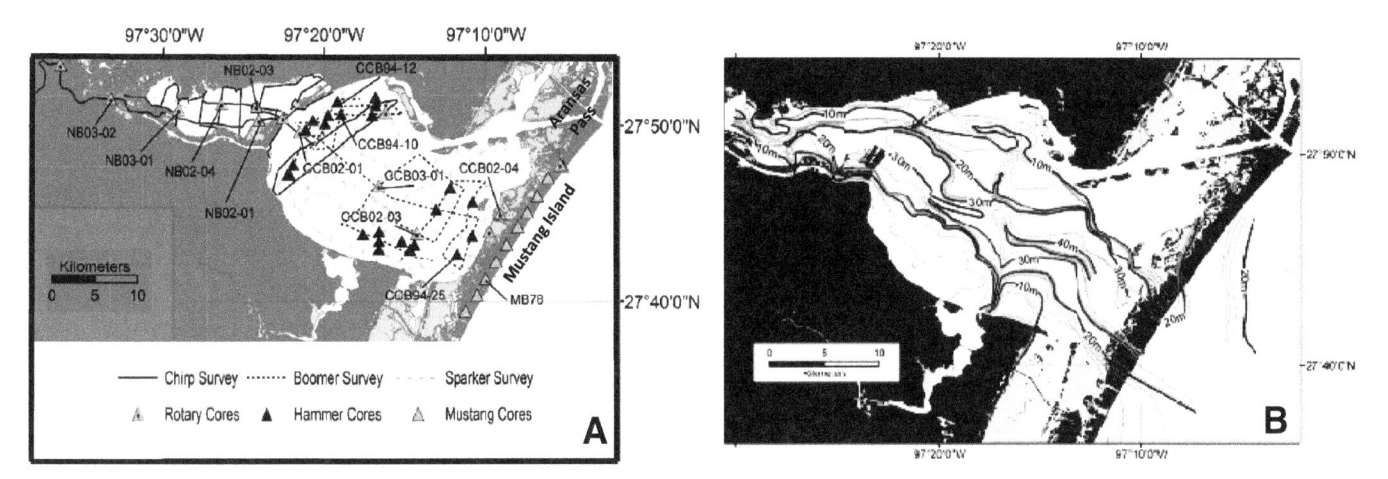

Figure 36. (A) Seismic lines and core locations used to study the evolution of Corpus Christi Bay are shown. (B) Isopach map of the Pleistocene surface shows Nueces incised valley within Corpus Christi Bay. Both maps were modified from Simms et al. (2008).

if ancestral Mustang Island was severely eroded, which removes it as an effective salinity barrier to the Gulf of Mexico.

Simms et al. (2008) interpreted a ca. 4.8 ka flooding event as indicating either a climatic change characterized by warmer and drier conditions compared to present, which occurred around this time, and/or to the flooding of a fluvial terrace. They further argued that the most recent flooding event (ca. 2.6 ka) resulted from a decrease in bay sediment delivery caused by a return to more mesic conditions, similar to those of the present climate. These interpretations are consistent with results from Ferguson et al. (2018b), who examined pollen records from the bay. Their results indicate decreased fluvial discharge to the bay at ca. 4.1 ka and ca. 2.2 ka, which they attributed to drier conditions at these times.

Rice et al. (2020) collected 38 cores of up to 15 m in length and 45 radiocarbon ages from the Nueces Delta in the upper reaches of Nueces Bay. These cores and ages confirmed the magnitude of the ca. 8.2 ka flooding event found in earlier studies but also revealed a strong influence of local climate changes on the upper reaches of the bay. Two periods of transgression occurred at ca. 5.0 ka and ca. 3.2 ka in which the delta retreated as much as 22 km within a few centuries (Rice et al., 2020). These periods of transgression, as well as intervening periods of progradation at 6.6 ka, 3.8 ka, and 2.2 ka, were linked to periods of Holocene climate change recorded by foraminiferal assemblages in a sediment core from Baffin Bay (Buzas-Stephens et al., 2014), which suggests that the shoreline retreated during dry/warm phases and prograded during wet/cooler phases.

In summary, central Texas bays appear to have been impacted by the same early Holocene sea-level events that are recorded in Galveston Bay, Sabine Lake, and Lake Calcasieu, which lends support to the argument that bay evolution was strongly influenced by punctuated sea-level events. These bays are also yielding a record of climate change in the region, in particular an increase in aridity during the late Holocene that altered the evolution of the bays during their later stages of formation.

Evolution of Coastal Barriers. Rodriguez et al. (2001) examined transects of sediment cores collected offshore of the eastern and central Texas Coast and noted along-strike variability in the thickness and stratigraphy of Holocene deposits (Fig. 38). They attributed this variability largely to differences in the antecedent topography of the Pleistocene surface on which these shoreface deposits and associated coastal barriers rest. Here, we present evidence that this variability is also due to differences in the volumes and timing of sand supply to the eastern and central Texas coasts.

Matagorda Peninsula. Matagorda Peninsula is a long, narrow peninsula that extends across the mouth of Matagorda Bay; the western portion of the peninsula is situated above the Lavaca River incised valley (Fig. 33). The peninsula averages 1.5 km wide and is mostly less than 2 m above sea level except for a narrow band of beach ridges that occurs near the peninsula's western end. Aerial photographs show that the peninsula is dominated by washover deposits, which indicates a recent history of landward retreat (Fig. 39).

Seismic profile MB02, which was collected along the bay side of the western part of the peninsula, shows bi-directional lateral accretion associated with a former inlet centered over the Lavaca valley and ~1.5 km landward of the current shoreline (Fig. 35). Drill core MB1–03 penetrated this unit, and radiocarbon dates from above and below indicate that the inlet was active between ca. 7.4 ka and ca. 6.7 ka (Fig. 34).

The first detailed investigation of Matagorda Peninsula was conducted by Wilkinson and Basse (1978) using a series of cores collected along the north side of the peninsula (Fig. 40). Cores on the west end of the peninsula sampled tidal deposits at the same location and depth as those imaged in seismic line MB02, which range in age from ca. 7.4 to ca. 6.7 ka (Fig. 35). This indicates that the shoreline was located ~1.5 km landward of its current location at that time. Above these tidal deposits, cores sampled gray and reddish-brown bay mud. The reddish-brown sediment color is characteristic of both the Brazos and Colorado rivers, whose drainage basins include extensive Permian and Triassic

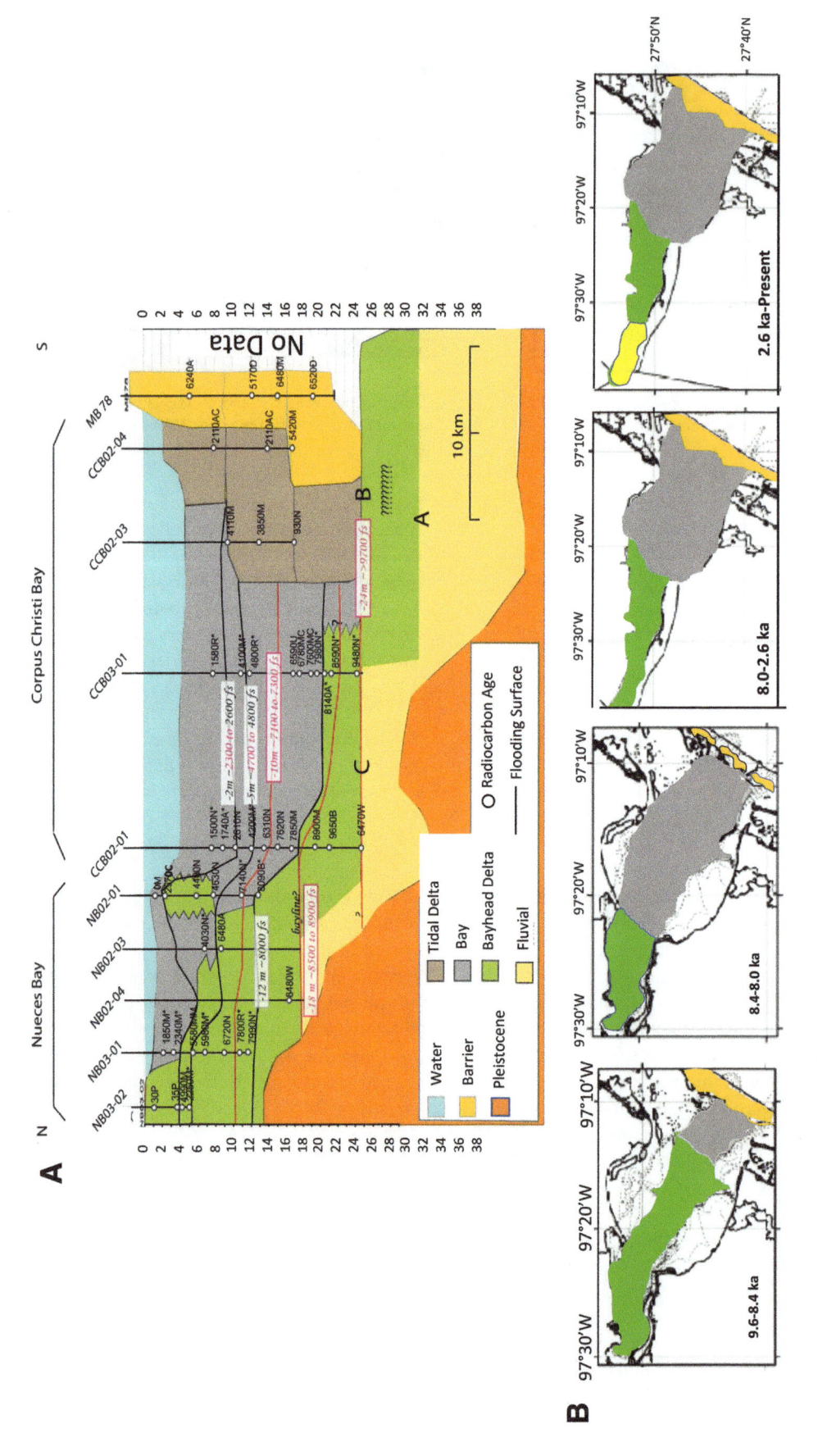

Figure 37. (A) North–south stratigraphic section for Corpus Christi Bay was compiled from seismic data and drill cores collected along the axis of the Nueces incised valley (see Fig. 36 for locations of drill cores). Also shown are flooding surfaces (fs) that mark episodes of landward-shifting bay facies, with radiocarbon dates used to constrain the ages of these surfaces (modified from Simms et al., 2008). (B) Paleogeographic maps depict changes in bay evolution based on data shown in the above cross section (modified from Simms et al., 2008).

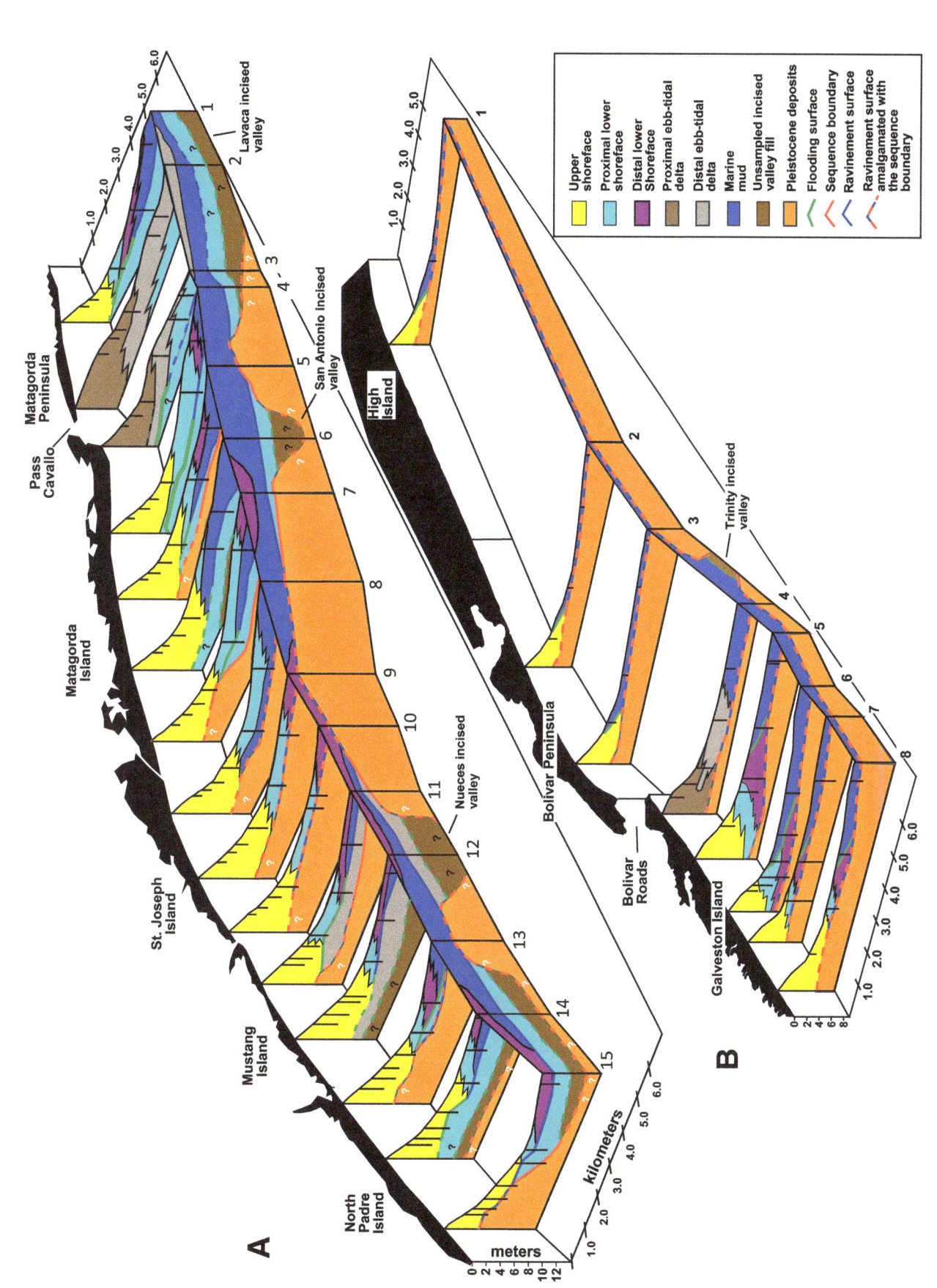

Figure 38. Fence diagram shows shoreface facies architecture offshore (A) central and (B) eastern Texas (modified from Rodriguez et al., 2001). Note the irregular relief on the Pleistocene surface, which reflects variable fluvial incision formed during the marine isotope stage (MIS) 2 lowstand; the main incised valleys are labeled (modified from Rodriguez et al., 2001). The thickest and most complete Holocene strata occur within these valleys.

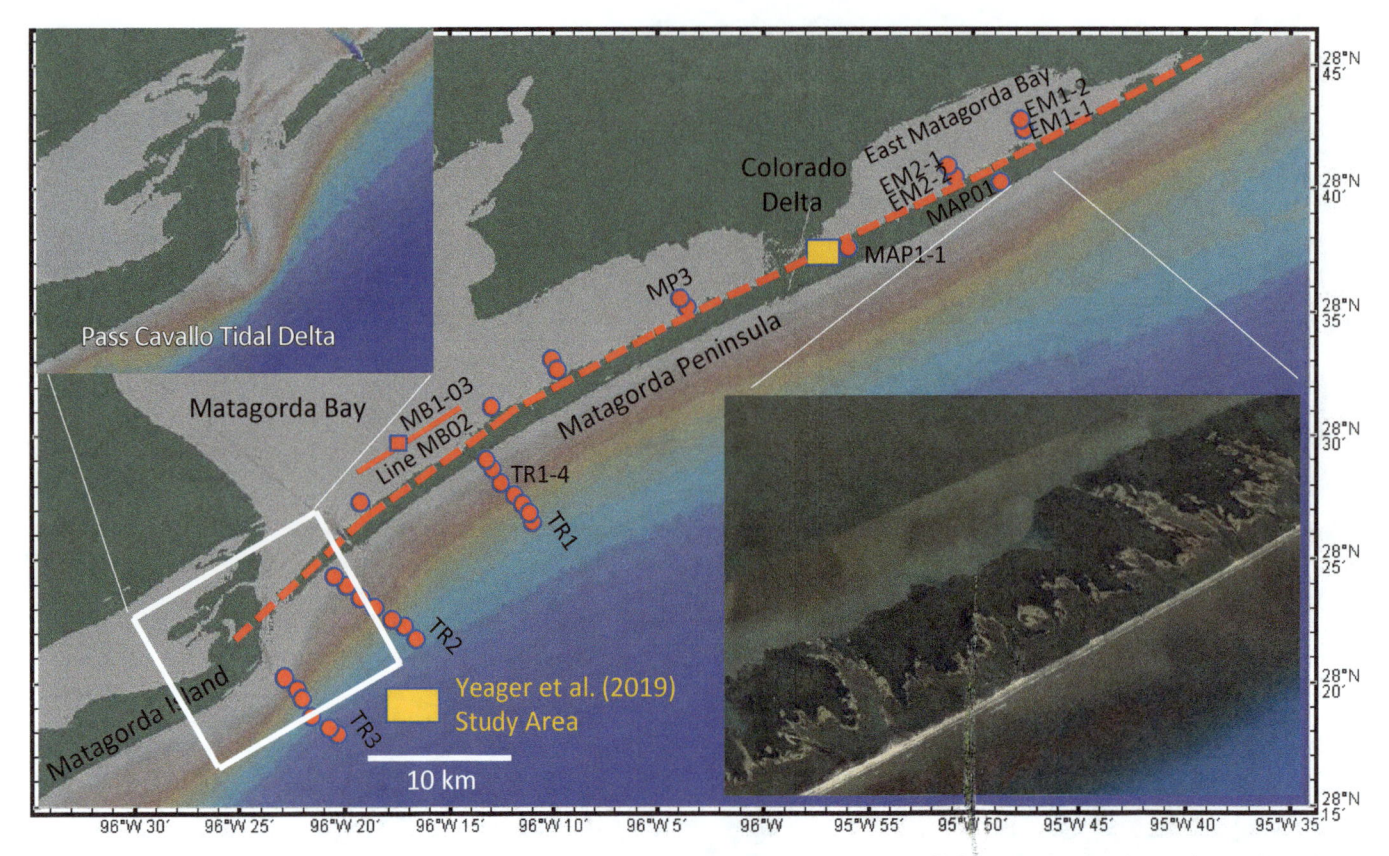

Figure 39. Locations of sediment cores (red circles) used to investigate the Holocene evolution of Matagorda Peninsula are shown. Also shown is the location of a study by Yeager et al. (2019), and enlarged images of a section of the peninsula show storm washover features (Google Earth) and the Pass Cavallo Tidal Delta (base map from GeoMapApp [www.geomapapp.org] / CC BY). The dashed red line shows the approximate location of the section compiled by Wilkinson and Basse (1978) shown in Figure 40.

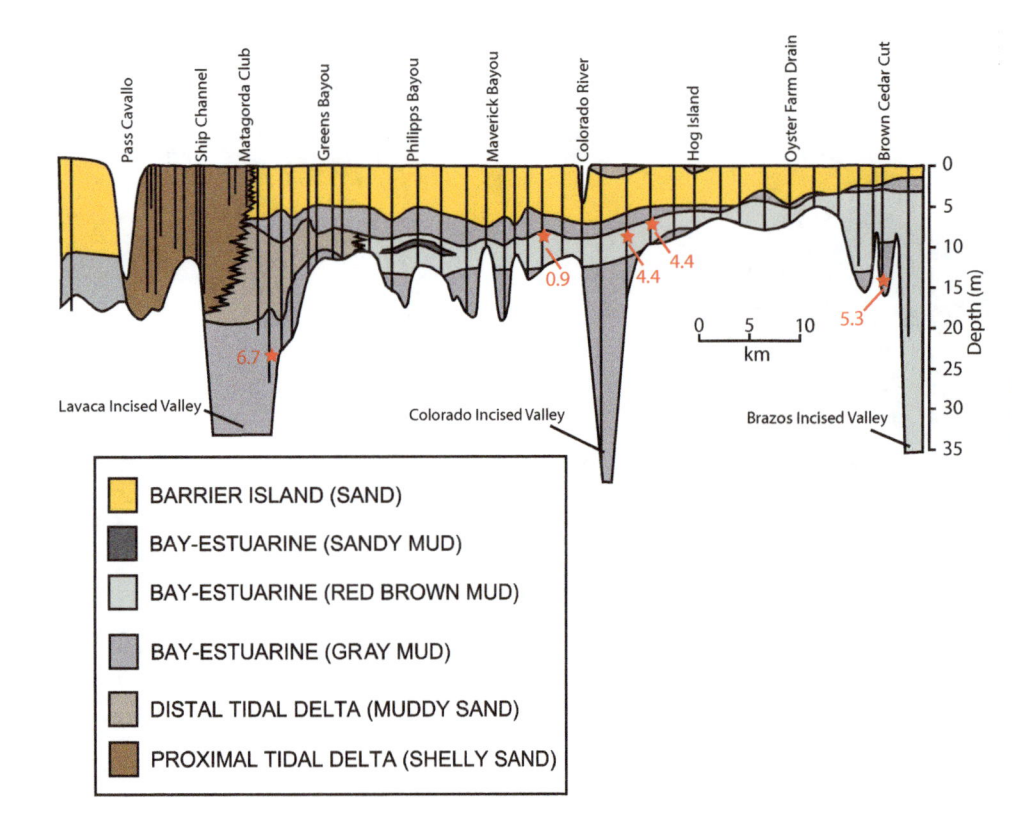

Figure 40. Shore-parallel cross section through Matagorda Peninsula with radiocarbon ages (in ka; red stars) was redrawn from Wilkinson and Basse (1978). See Figure 39 for section location.

redbeds. This color change likely records the avulsion of the river from the center of its incised valley to a more western location. The radiocarbon ages from the overlying bay deposits span 5.3 ka to 4.4 ka, with one anomalous age of 0.9 ka.

Yeager et al. (2019) collected multiple vibracores along three transects just east of the Colorado River (Fig. 39). They acquired a total of 49 radiocarbon ages from these cores that can be used to better constrain the ages of the key stratigraphic surfaces observed by Wilkinson and Basse (1978). The oldest unit sampled in their cores was the reddish-brown bay mud, whose age was derived from four radiocarbon ages that range from ca. 4.8 ka to ca. 4.4 ka. These ages are consistent with those acquired by Wilkinson and Basse (1978) for this unit. Yeager et al. (2019) acquired 13 radiocarbon ages from a gray bay mud unit above the reddish brown mud unit. These ages cluster between ca. 4.3 ka and ca. 2.4 ka, except for one anomalous date of 5174 cal yr B.P., which is assumed to be from re-worked material.

A prominent shell hash unit near the top of the upper bay mud unit is interpreted as a transgressive surface formed by landward migration of Matagorda Peninsula. Yeager et al. (2019) obtained 27 radiocarbon ages from this shell hash that cluster between ca. 3.1 ka and ca. 2.0 ka. Ages occur in stratigraphic order in cores in which more than one age was obtained. The timing of this late Holocene transgression is generally consistent with results from Matagorda Bay, which indicate that the bay experienced widening and increased marine influence during this time (Fig. 34).

The Wilkinson and Basse (1978) profile through the Matagorda Peninsula shows an upper sand unit that averages 6 m thick and was interpreted as barrier island sand (Fig. 40). The cores acquired by Yeager et al. (2019) were collected near the center of the peninsula and sampled an average of 0.70–1.3 m of barrier sand resting on bay deposits. We collected two cores (MAP01 and MAP1-1; Fig. 39) from the foreshore and upper shoreface not far from their study area that sampled between 1 m and 1.5 m of sand resting on bay mud. These findings suggest that the upper barrier unit identified by Wilkinson and Basse includes tidal deposits and/or washover deposits and that the barrier sands are less than ~2.5 m thick, including barrier topography.

Yeager et al. (2019) obtained two radiocarbon ages from barrier sands of 1727 cal yr B.P. and 1262 cal yr B.P. We collected 11 vibracores behind Matagorda Peninsula that further constrain the transgressive phase of barrier development as recorded by the onset of washover deposition. Four cores (cores EM1-1, EM1-2, EM2-1, and EM2-2; Fig. 39), collected behind the eastern part of the peninsula, sampled proximal washover sands that vary in thickness from 100 cm in the most distal cores to 190 cm in the most proximal cores. The sampled succession is composed of, from bottom to top, gray bay mud/proximal washover sands/distal washover muddy sand/sandy mud. This succession indicates a decrease in the bayward extent of washover deposition through time. Core EM 2-1 sampled bay mud at the base of washover sands that yielded an age of 2220 cal yr B.P. Core EM 1-2 sampled a similar succession and yielded a radiocarbon age of 1680 cal yr B.P. from bay mud beneath washover deposits. Core EM 1-1 yielded the youngest age of bay muds: 1160 cal yr B.P. Core MP3, collected west of the Colorado Delta (Fig. 39), did not penetrate the base of the washover deposits, and a shell from 103 cm depth in the core within washover sands yielded a modern age. These combined results indicate that deposition of proximal washover deposits behind Matagorda Peninsula has been occurring since ca. 1.3 ka to ca. 1.2 ka, which is generally consistent with results from Yeager et al. (2019) indicating deposition of the upper transgressive barrier sand unit since ca. 1.7 ka.

Rodriguez et al. (2001) and Odezulu et al. (2020) studied cores from two offshore transects (TR1 and TR2) that extend from the upper shoreface to 5 km offshore of the west end of Matagorda Peninsula (Fig. 39). Both transects sampled a stratigraphic succession indicating an early phase of shoreface progradation, followed by a transgressive event marked by onlapping marine mud, and ending with a later phase of shoreface progradation (Fig. 41). In transect 1, the older shoreface progradation extended over 5 km from the current shoreline (Fig. 41A). This phase of shoreface progradation ended with a transgressive event that is marked by ~5 km of marine mud onlap (transgressive ravinement surface [TRS] in Fig. 41A). The more recent phase of progradation resulted in shoreface deposits extending several kilometers seaward of the current shoreline. No reliable radiocarbon ages were obtained from these cores; however, the sequence of regressive and transgressive events recorded by this core transect is consistent with those recorded by seismic profiles and drill cores from Matagorda Bay (Maddox et al., 2008) and results from onshore investigations (Wilkinson and Basse, 1978; Yeager et al., 2019). These combined data indicate that a tidal inlet and associated tidal delta existed near the current shoreline between ca. 7.4 ka and 6.7 ka, followed by progradation that led to tidal deposits being buried beneath bay mud. This mid-Holocene event is correlated to the lower progradational unit in TR 1 (Fig. 41A). This was followed by transgression that is manifest offshore by the onlap of marine mud and culminated with the peninsula shifting to its current location by ca. 1.7 ka. Offshore cores indicate that the shoreface has experienced progradation since the peninsula reached its current location, although this overlaps with the age of washover deposits (1245–1160 cal yr B.P.).

The Pass Cavallo tidal inlet and delta separates Matagorda Peninsula from Matagorda Island (Fig. 39). Core transects TR 2 and TR 3, located offshore of Pass Cavallo, sampled ebb tidal delta deposits composed of relatively coarse (170–270 µm) sand with shell debris layers (Fig. 38). These deposits extend nearly 3 km offshore and perhaps explain the shoreface progradation recorded in offshore core transect TR1. This further indicates that the Pass Cavallo tidal inlet and tidal delta complex has existed near its current location since ca. 1.7 ka, when the shoreline reached its present location, and that only minor landward migration of Matagorda Peninsula has occurred since this time. Historical shoreline migration records indicate that Matagorda Peninsula has been retreating landward at a rate of 0.9 m/yr (Paine et al., 2021).

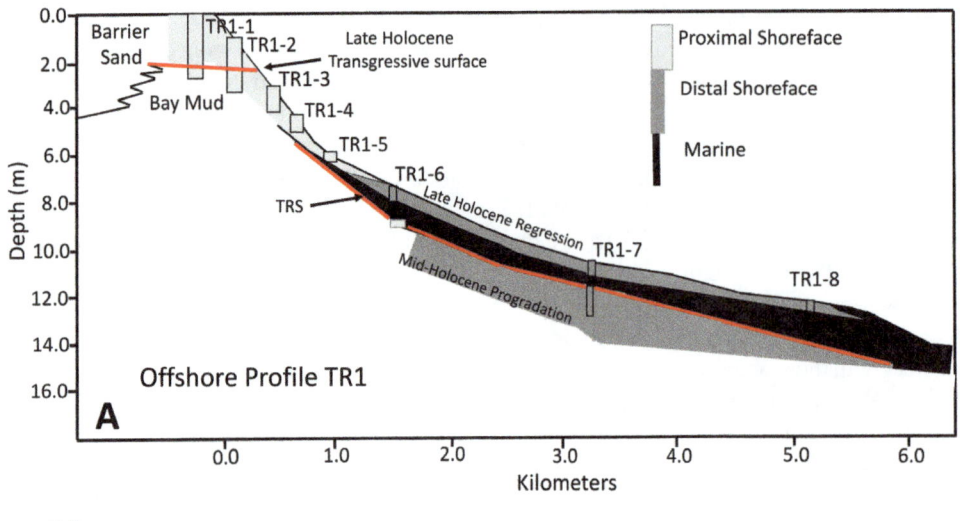

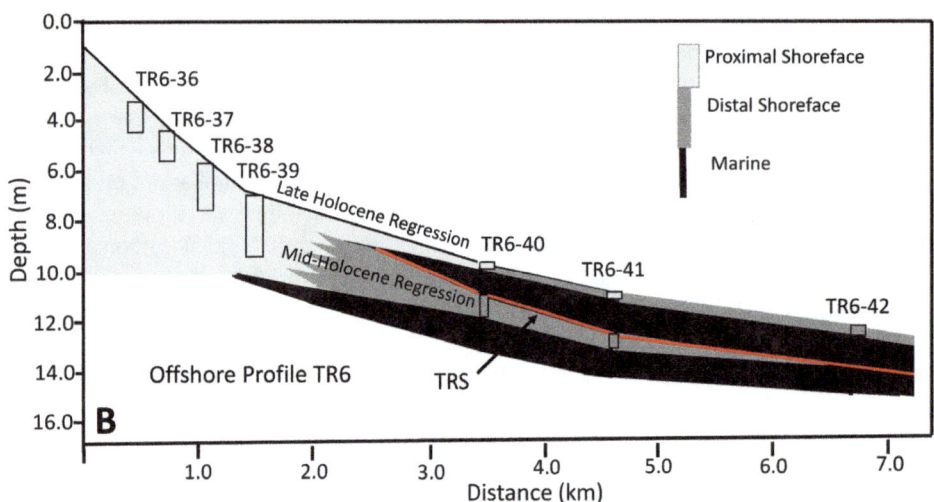

Figure 41. Stratigraphic sections from core transects TR-1 (A) and TR-6 (B) off Matagorda Island and Matagorda Peninsula show two episodes of shoreface progradation, one during the mid-Holocene and another during the late Holocene, separated by a transgressive surface (shown by red line) and onlap by a marine mud unit (modified from Odezulu et al., 2020). See Figures 39 and 44 for profile locations. TRS—transgressive ravinement surface.

Matagorda Island and San Jose Island. Matagorda Island and San Jose Island differ from Matagorda Peninsula in their pronounced ridge and swale topography (Figs. 42B–42C). These are among the highest beach ridges on the Texas Coast, with the more prominent ridges typically having heights of 3–4 m. These ridges extend across Cedar Bayou (Fig. 42A), which is the geographic boundary between the two islands. Also, sediment cores collected on both the bay side and offshore of Cedar Bayou did not sample tidal delta deposits (Fig. 38, Transect 7), which indicates that the bayou was never an active tidal inlet. Thus, the two islands are assumed to have evolved as one ~90-km-long barrier. The beach-ridge fields of both islands are bordered on their bay sides by washover features (Fig. 42C). The width of the beach ridge field is wider on Matagorda Island, on average 1.9 km at the east end of the island, as compared to 0.6 km on San Jose Island. The washover features on the bay side of the islands appear to pre-date the beach ridge fields based on the relatively sharp boundary that separates these features, evidence for drowning and degradation of washover fans, and the fact that the ridges are breached by storm-surge channels only at the narrow western end of San Jose Island. This geomorphic setting indicates that an early transgressive phase of barrier development was followed by progradation.

The east end of Matagorda Island lies above the western margin of the Lavaca incised river valley, and the central part of the island straddles the San Antonio incised valley (Fig. 38). A shore-parallel section of cores collected through Matagorda Island shows that barrier sand averages 12 m in thickness (Fig. 43), which is about twice the thickness of sand on Matagorda Peninsula (Wilkinson, 1975). The thickness of San Jose Island is not known, but it is assumed to be thinner than Matagorda Island since it rests on a Pleistocene high (Fig. 38). Vibracores collected on the western end of the island (SJ1–SJ3; Fig. 44) bottomed out in barrier sands that are more than 5 m thick.

The stratigraphic section of Matagorda Island, as described by Wilkinson (1975), is similar to that of Matagorda Peninsula and is composed mainly of a transgressive sequence composed

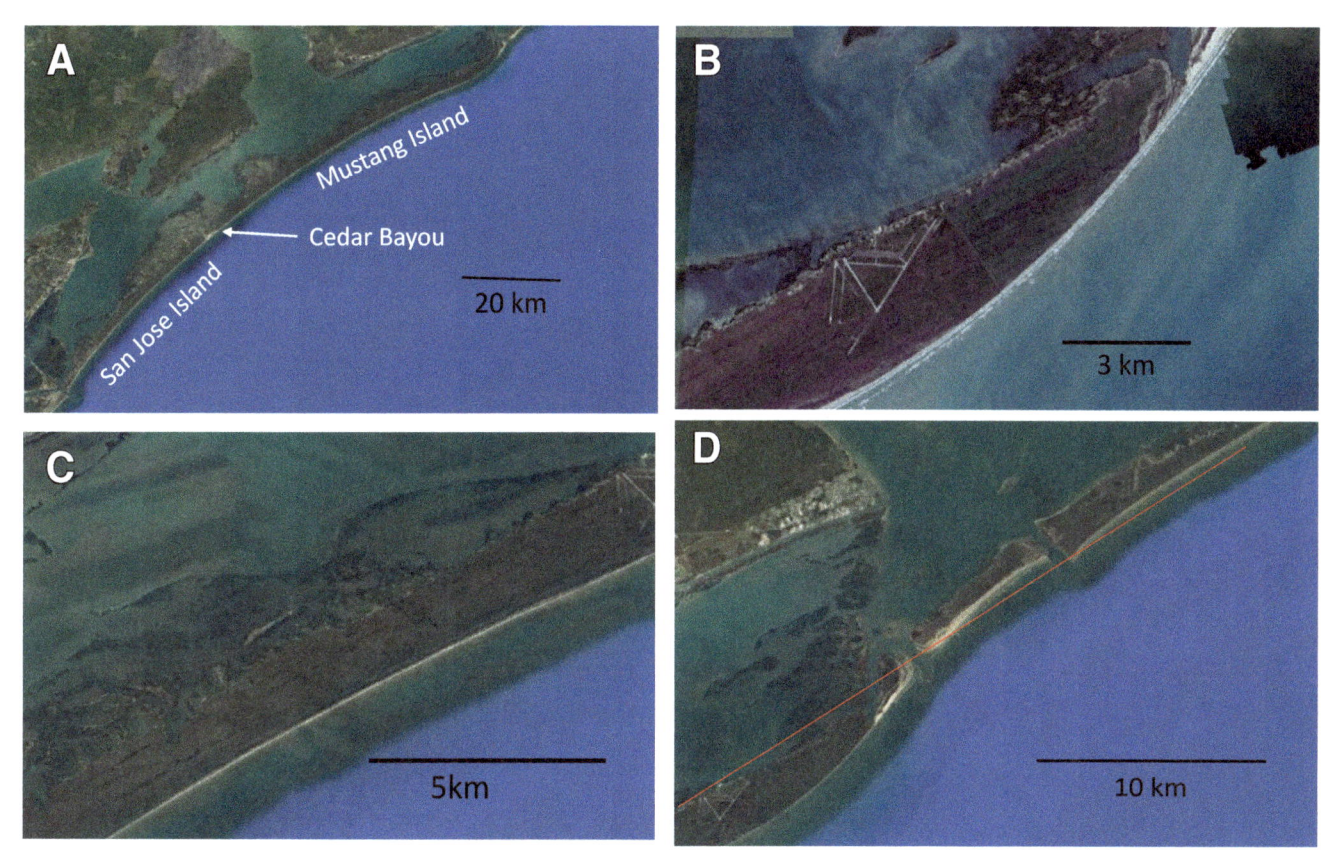

Figure 42. Google Earth images show (A) Mustang Island and San Jose Island separated by Cedar Bayou, (B) truncated beach ridges at the east end of Matagorda Island, (C) washover fans behind Matagorda Island, and (D) offset between Matagorda Peninsula and Matagorda Island highlighted by red line.

of, from bottom to top, Pleistocene Ingleside sand resting on a soil horizon (MIS 5e deposits)/fluvial deltaic sand that fills the San Antonio River incised valley/bay-estuarine mud/barrier island sand (Fig. 43). The regressive and transgressive phases of barrier evolution that are recorded by the geomorphology of the island occur within the upper barrier sand unit. Radiocarbon dates are lacking for Matagorda Island, but similarity of its stratigraphy to that of Matagorda Peninsula indicates that the ages of the bay and barrier units can be inferred using the well-constrained ages for the peninsula. Based on this comparison, the bay mud unit ranges from ca. 4.8 ka to ca. 2.0 ka, with barrier sands having been deposited after ca. 2.0 ka.

We collected eight vibracores behind Matagorda Island and San Jose Island in an attempt to obtain radiocarbon ages that would better constrain the timing of transgression of these barriers (Fig. 44). Only one reliable age of 2252 cal yr B.P. was obtained from washover deposits behind Matagorda Island from core MI6. Three other samples yielded radiocarbon dead ages, which indicate that they are from reworked material. Core SJ4 was collected from a washover fan on the bay side of San Jose Island and yielded an age of 2557 cal yr B.P. The combined results indicate that both islands were experiencing transgression and washover during this time interval.

A transect of three vibracores from the western end of San Jose Island (cores SJI1–SJI3; Fig. 44) was acquired within the beach ridge field and was aimed at constraining the timing of the late stage of island progradation. Core SJI1 was collected from the landward-most beach ridge on the island, core SJI3 from near the center of the beach ridge complex, and core SJI2 was collected from the youngest beach ridges near the current gulf shoreline. Two samples from core SJI1 yielded ages of 1595 cal yr B.P. and 1560 cal yr B.P. A single sample from SJI3 yielded an age of 1342 cal yr B.P., and a modern date was obtained for the youngest beach ridge (SJI2). The oldest part of the barrier has not been sampled. These combined ages indicate that the late stage of island progradation began at least at ca. 2.3 ka (shift from overwash to beach ridges) and continued until modern time. The island has prograded over 1 km since ca. 1.4 ka at an average rate of 0.8 m/yr.

Eight core transects collected offshore of Matagorda Peninsula to San Jose Island (Fig. 38) were studied by Rodriguez et al. (2001) and Odezulu et al. (2020). Because Matagorda Island rests above the San Antonio River incised valley, core transects offshore of the island sampled a thicker, more complete record of shoreface evolution than occurs off San Jose Island, which sits above a Pleistocene high (Fig. 38). These transects show a

48   Anderson et al.

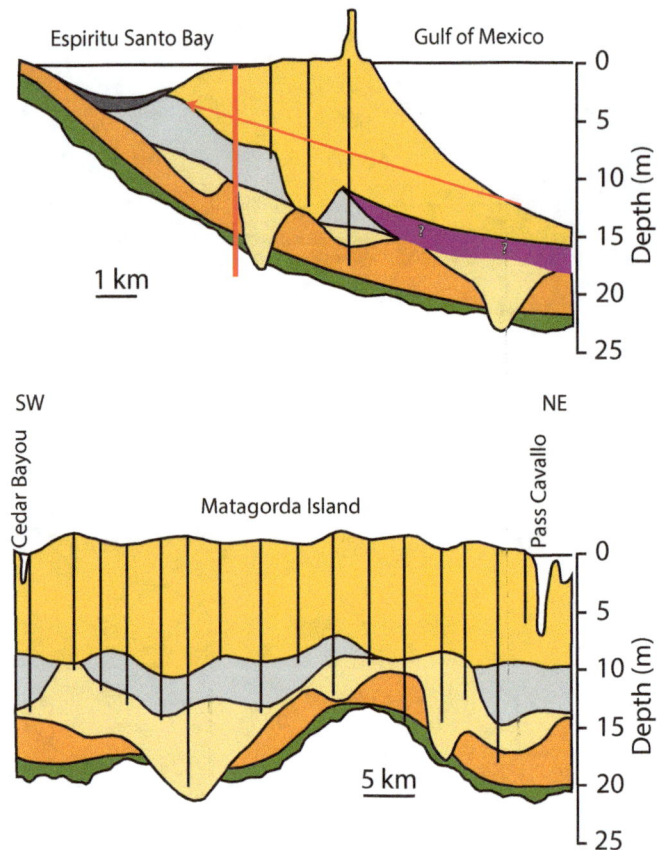

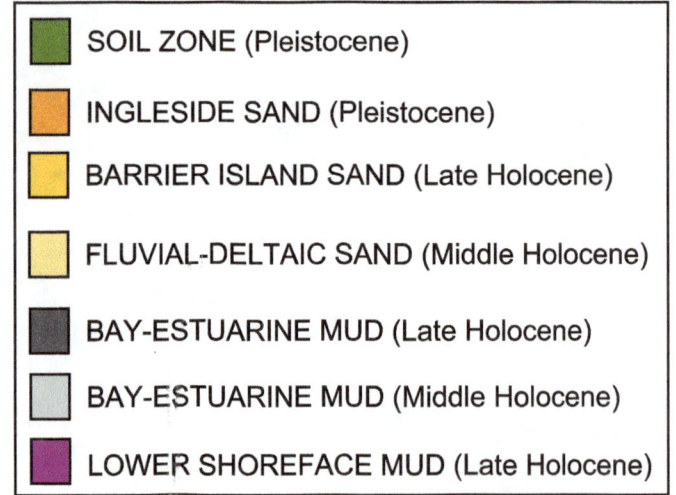

Figure 43. (Top) cross section through Matagorda Island and (bottom) shore-parallel stratigraphic section through the island (redrawn from Wilkinson, 1975). The heavy red line in the upper section shows the approximate location of the red line in Figure 42D, which illustrates the approximate location where the transgressive surface cuts through Matagorda Peninsula, resulting in thinner barrier and shoreface sands on the peninsula relative to Matagorda Island. The thin red line shows the approximate location of the transgressive surface. Location of top panel is shown by red line in Figure 44 through Matagorda Island.

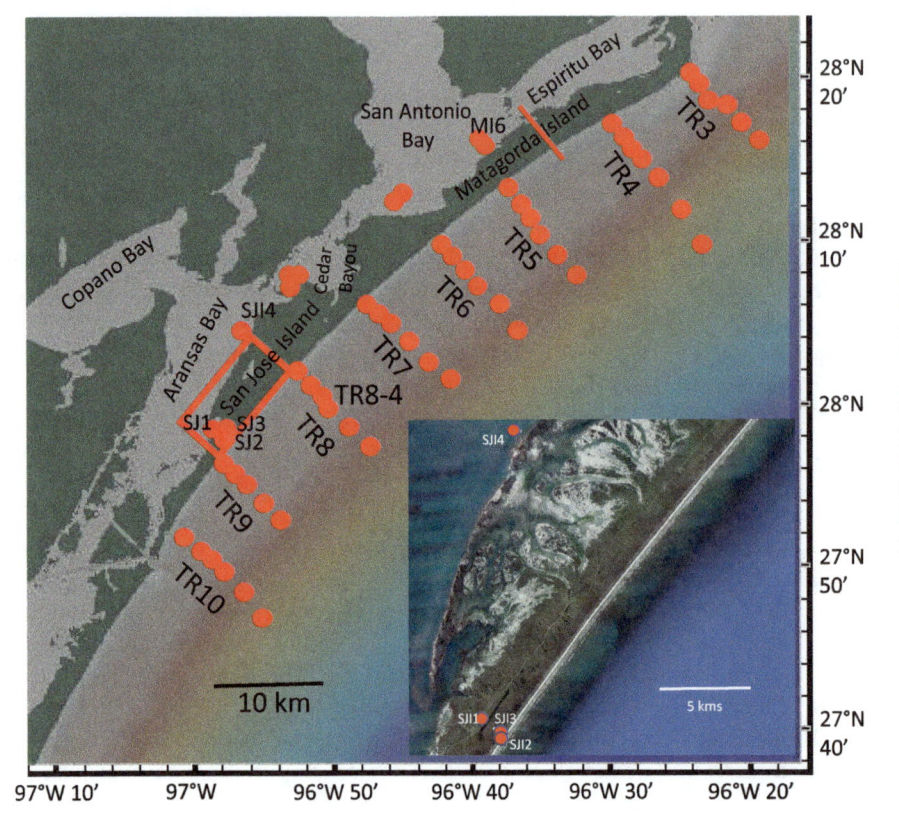

Figure 44. Locations of sediment cores (red circles) used to investigate the Holocene evolution of Matagorda Island and San Jose Island (base map from GeoMapApp [www.geomapapp.org] / CC BY). Transect numbers are the same as those shown in Figure 38. Red box indicates location where cores from San Jose Island were collected (base map from Google Earth). Red line through Matagorda Island shows approximate location of the top panel of Figure 43.

stratigraphic architecture similar to that observed in transects TR1-1 and TR1-2 collected off Matagorda Peninsula, with two regressive events separated by a major transgressive event (Fig. 41). The exceptions are profiles TR8, TR9, and TR10, which also occur above a Pleistocene high and sampled condensed sections that do not record the earlier regressive event (Fig. 38). These regressive and transgressive events indicate shoreface migration on the order of 2–3 km. Given the magnitude of these events, it is reasonable to assume that they coincide with the transgression and progradation recorded onshore.

Wilkinson (1975) observed that the ridge and swale topography of Matagorda Island is truncated near the eastern end of the island (Fig. 42B), which indicates a recent period of erosion of the island. He suggested that this erosional event was responsible for reducing Matagorda Peninsula to its narrower and thinner condition relative to Matagorda Island. If we assume that the two barriers were initially aligned with one another, this implies that the shoreline of western Matagorda Peninsula has migrated landward ~2.0 km more than that of Matagorda Island, which has removed beach ridges and left only a thin, transgressive remnant of the barrier (Fig. 42D). This is consistent with results from offshore core transects that show a relatively thin (<2.0 m thick), narrow (~1 km) upper shoreface succession off Matagorda Peninsula compared to offshore Matagorda Island (Fig. 41).

Historical shoreline change records indicate that the Matagorda Island shoreline has been relatively stationary while the Matagorda Peninsula and San Jose Island shorelines have migrated landward at rates of −0.9 m/yr and −0.8 m/yr, respectively (Paine et al., 2021). The sand eroded from Matagorda Peninsula contributed to the growth of the Pass Cavallo Tidal Delta. The Pass Cavallo Tidal Delta has not been studied in detail, so its sand volume remains unknown. Based on bathymetric data, the combined flood and ebb tidal deltas cover an area of ~70 km^2. If we assume that its sand component averages 2 m in thickness over this area, the tidal delta contains as much as 140 × 10^6 m^3 sand. Based on our sediment cores from Matagorda Bay, we estimate a roughly equal amount of sand storage in washover deposits. We do not have data that allow a volume estimation for offshore storm beds, but similar to conditions at San Luis Pass Tidal Delta, these combined depocenters are considered a significant sink for sand eroded from Matagorda Peninsula and the east end of Matagorda Island during the most recent transgressive event.

Mustang Island and North Padre Island. San Jose Island and Mustang Island are currently separated by Aransas Pass, but work by Simms et al. (2008) indicates that the main tidal inlet for Corpus Christi Bay was situated near the central and western end of the bay throughout most the island's history. Core transects collected offshore Aransas Pass (transects 10–11, Fig. 38) did not sample late Holocene ebb-tidal delta deposits, so the current tidal inlet is considered a modern feature. Historical charts show an ebb tidal delta and associated tidal inlet that migrated westward nearly 2 km from 1866 until the ship channel and associated jetties were constructed; work began in 1885 and was completed in 1919. This indicates an inlet migration rate of more than 1 km per century. Since the jetties were constructed, sand has accumulated mainly on the south side of the jetties, which reflects the predominantly northward flow of longshore currents in the region.

Both Mustang Island and North Padre Island are characterized by landscapes that lack beach ridges, and washover fans are highly degraded. The two islands are geographically separated by Packery Channel, which is a small storm surge channel that is not permanently connected to the Gulf. Sediment core transects collected offshore of both islands show similar stratigraphic records, and cores collected offshore of Packery Channel did not sample ebb tidal delta deposits, which suggests that this channel is a modern feature (Fig. 38, Transects 12 and 13). Thus, Mustang Island and North Padre Island are assumed to have evolved as one continuous barrier system. A Pleistocene high does occur between Mustang Island, which sits above the Nueces valley, and North Padre Island, which sits above the Baffin Bay valley (Fig. 38). This explains the significant differences in the thickness of Holocene sediments beneath the two islands.

Mustang Island has been the site of several investigations, but work on North Padre Island has been restricted to the northern part of the island. However, offshore core transects do provide an opportunity for offshore stratigraphic comparisons of the two islands (Fig. 45). Shideler (1986) used platform borings to construct a cross section through Mustang Island (Fig. 45A). This section shows that the island is upwards of 20 m thick where it sits above the Nueces River incised valley. Only a few radiocarbon ages were acquired from a single core (core EE; Fig. 45B), which indicates that the barrier sands are mostly younger than 1470 cal yr B.P. at this location. Shideler's work was followed by two field campaigns that focused on the barrier's evolution (Simms et al., 2006; Garrison et al., 2010). Simms et al. (2006) used a shore-parallel transect of long cores, most of which penetrated to the Pleistocene surface, to construct a shore-parallel section for Mustang Island (Fig. 45C). These cores sampled up to 32 m of well-sorted, medium to fine sand interbedded with shell hash layers, which is interpreted as mostly tidal deposits within stacked tidal channels.

Garrison et al. (2010) collected a 3-km-long ground-penetrating radar (GPR) profile across the north side of North Padre Island that shows landward-dipping reflectors on the bay side (1 km inland of the bay shore) of the island and seaward-dipping reflectors on the gulf side (2 km inland from the gulf shoreline; Fig. 46). The landward-dipping reflectors are up to 10 m thick, which suggests that they are flood tidal delta deposits that filled a tidal incision. These landward- and seaward-dipping reflectors are truncated by an irregular erosion surface with several channels that cut as much as 6 m into the island. No cores or radiocarbon ages were collected, so the only age constraints for these GPR units were inferred from Shideler's core EE, which was collected nearly 30 km to the east. We acquired two GPR profiles from the west end of Mustang Island, one shore-parallel profile, and another profile that extends across the island from the beach and overlaps with the Simms et al. (2006) data set (Fig. 45C). Although these lines image only the

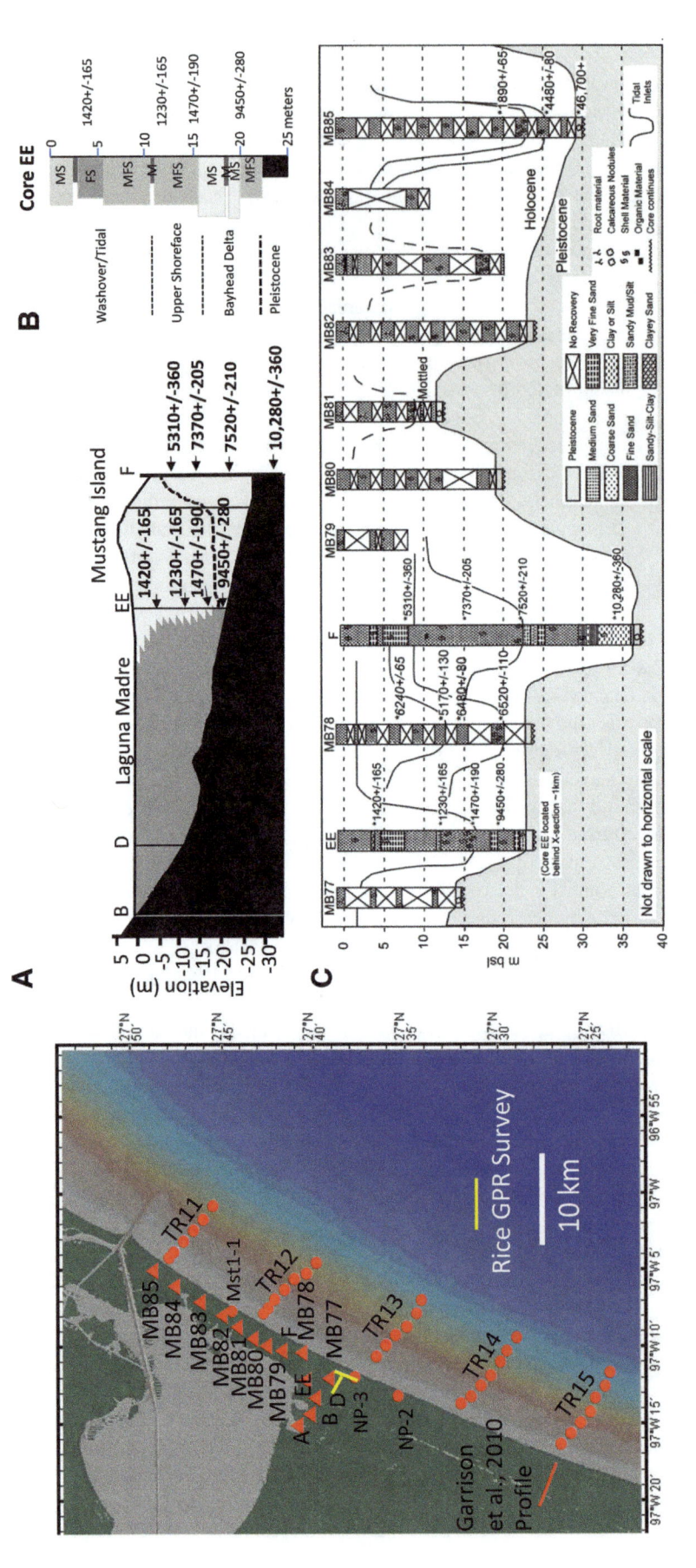

Figure 45. (Left) Locations of ground penetrating radar (GPR) profiles (red line is Garrison et al., 2010, profile and yellow lines are unpublished Rice University lines) and sediment cores (red circles) used to investigate the Holocene evolution of Mustang Island and North Padre Island are shown (base map from GeoMapApp [www.geomapapp.org] / CC BY). Transect numbers are the same as in Figure 38. (A) Northeast–southwest stratigraphic section for Mustang Island with radiocarbon ages. Depositional environment interpretations and radiocarbon ages for core F (Simms et al., 2006) and for core EE (from Shideler, 1986) (expanded in panel B). (C) Shore-parallel stratigraphic section through Mustang Island (from Simms et al., 2006).

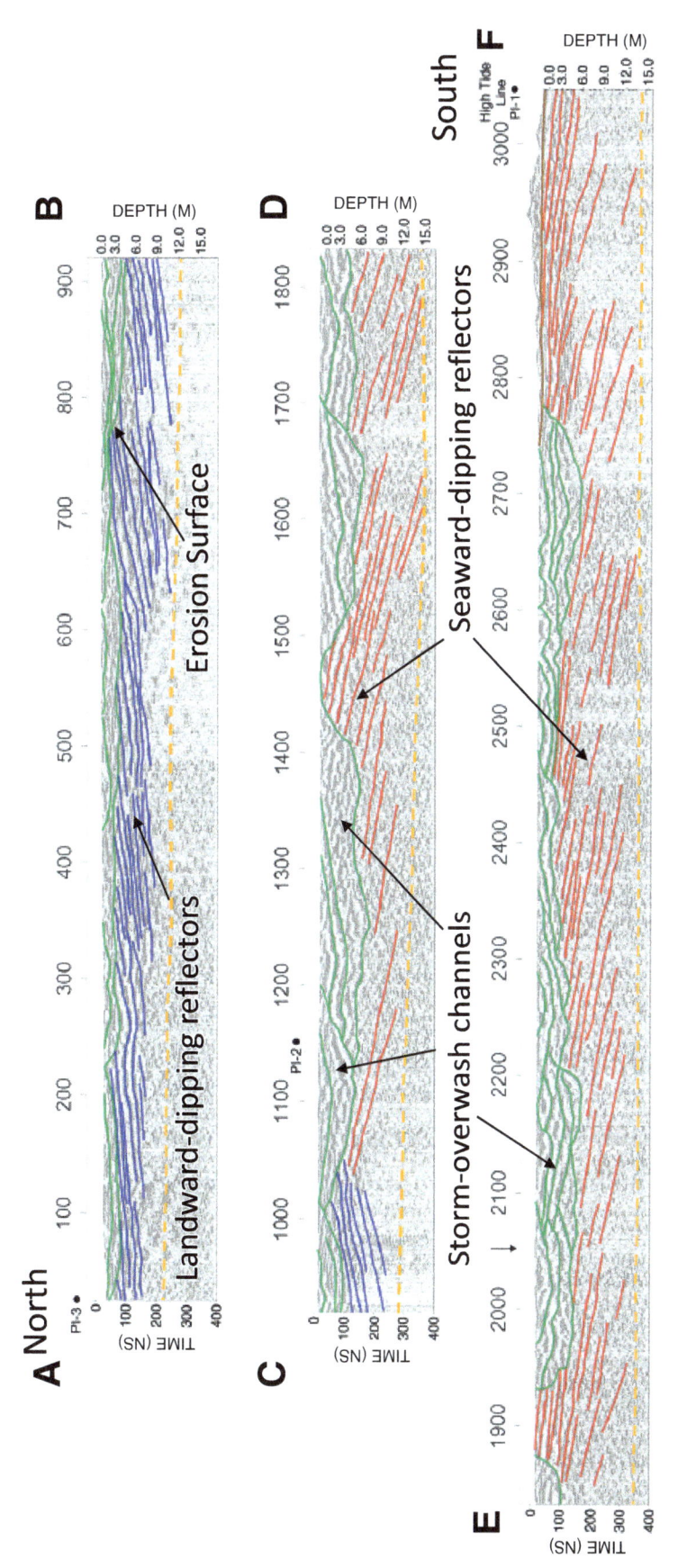

Figure 46. Interpreted ground-penetrating radar profile across North Padre Island is shown (modified from Garrison et al., 2010; republished with permission of SEPM [Society for Sedimentary Geology]; permission conveyed through Copyright Clearance Center, Inc.). See Figure 45 for profile location.

upper few meters of the section, they provide context for our onshore sediment cores.

Radiocarbon ages acquired by Simms et al. (2006) indicate that Mustang Island has been located near its current location since at least ca. 7.5 ka but possibly as early as 9.6 ka (Fig. 45C). The latter age is based on a single age from core EE of Shideler (1986), and the former is based on two ages of ca. 7.5 ka and three more in the 6–7 ka range (Simms et al., 2006). This is consistent with results from Corpus Christi Bay that show a major flooding event at ca. 8.2 ka (Simms et al., 2008; Ferguson et al., 2018a). Drill cores collected within Corpus Christi Bay indicate that the island thins abruptly to the west (Fig. 37) as shown in Figure 45B. Drill cores MB80 through EE were collected from the western part of the island and are located over the Nueces incised valley, which is up to 35 m deep at this location. These cores sampled mostly medium sand with thin shell beds. Shells from these cores yielded radiocarbon ages with anomalously young ages for their relatively deep depths, so they were interpreted as channels filled with mostly tidal inlet deposits (Simms et al., 2006). Simms et al. (2008) identified flood tidal delta deposits in the southwestern part of Corpus Christi Bay whose age extends back to 6520 cal yr B.P.

Cores MB85 and EE sampled expanded middle to late Holocene sections along the eastern and western margins of the Nueces valley. Shideler's core EE was collected within the western channel and yielded a radiocarbon age from bayhead delta deposits near the base of the channel of 9450 cal yr B.P. (Fig. 45B). Most of the channel is filled with sand that Shideler interpreted as an upper shoreface deposit, suggesting that the contact between these units is a transgressive surface. The shoreface deposits are overlain by a unit that is interpreted as washover deposits, which suggests a regressive phase of barrier evolution. Radiocarbon ages indicate that this regression was underway by 1470 cal yr B.P. (Fig. 45C).

Five core transects collected offshore of Mustang and North Padre Islands provide additional details about these barrier's evolution (Rodriguez et al., 2001; Odezulu et al., 2020). Transects TR-11 and TR-12 were collected offshore of Mustang Island (Fig. 38). The stratigraphic successions recovered in both core transects record two episodes of shoreface progradation marked by two fining offshore sand units, from fine to very fine sand, separated by marine mud (Fig. 47). The oldest regressive unit rests on sand with shell hash, which is interpreted as an ebb tidal delta deposit based on its coarser grain size relative to the barrier sands (Odezulu et al., 2020). These ebb tidal delta and shoreface sands extend 5 km and 6 km seaward of the current shoreline, indicating significant progradation. A single radiocarbon age from the base of the shoreface sand in core TR12-83 indicates that progradation was underway by 7340 cal yr B.P. (Fig. 47). This age is consistent with results from Shideler (1986) and Simms et al. (2006) indicating that the barrier was located near its current location by this time. Odezulu et al. (2020) argue that this phase of progradation resulted from an increase in the flux of sand to the central Texas Coast from erosion of the Colorado and Rio Grande deltas, which existed as prominent delta headlands during the middle Holocene.

Marine mud directly overlies the mid-Holocene progradational unit and records a transgressive phase of barrier evolution (Fig. 47). Radiocarbon ages indicate that this transgressive event was underway by 4850 cal yr B.P. and ended by 1560 cal yr B.P. (Fig. 47). We suggest that this transgression resulted in the landward-dipping foresets, which are interpreted as washover and flood tidal deposits, imaged in Garrison et al.'s (2010) GPR profile. Odezulu et al. (2020) argue that this phase of barrier evolution resulted from expansion of the Texas Mud Blanket during this time interval. This transgressive event was followed by barrier progradation that is recorded in the GPR profile as seaward-dipping reflectors (Fig. 46). Three cores from the modern beach of western Mustang Island and northern North Padre Island targeted this progradational unit and a younger surface unit characterized by cross-cutting channels. These cores recovered between 1.5 m and 2.0 m of beach and foreshore deposits resting on upper shoreface deposits, thus yielding additional evidence for a recent phase of barrier progradation. In our offshore cores, this final stage of shoreface progradation is marked by a thin (< 2 m) shoreface unit that extends ~3 km seaward of the modern shoreline (Fig. 47). A radiocarbon age of 1560 cal yr B.P. from below the upper sand unit in core TR14-95 indicates that this most recent progradation occurred soon after the transgression that ended at ca. 1.6 ka. Three radiocarbon dates from the youngest progradational unit range from 0.5 ka to modern. This is consistent with results from Shideler (1986) indicating that the more recent growth of the island occurred after ca. 1.5 ka. Historical shoreline migration rates for Mustang Island indicate that the island is migrating landward at a rate of 0.3 m/yr (Paine et al., 2021), which is a departure from the progradation that has occurred since ca. 1.5 ka.

In summary, with the exception of Matagorda Peninsula, the coastal barriers of central Texas are the thickest and most robust barriers of the LaTex Coast. This is a reflection of the abundant sand supply to these barriers from erosion of the offshore Colorado and Rio Grande deltas and of the convergence of longshore currents that delivered this sand to these barriers. Their evolution was closely linked to the evolution of the Texas Mud Blanket, which buried offshore sand sources.

The South Texas Coast

The modern south Texas Coast is occupied by Baffin Bay, Laguna Madre, and South Padre Island (Fig. 1). The Rio Grande Delta dominates the offshore bathymetry, and reworking of the delta has played a key role in coastal evolution as a source of sand and as an influence on wave climate. South Texas has a semi-arid climate, with annual precipitation of <50 cm/yr (Fig. 3). The semiarid setting supports a sparse vegetation cover, relative to areas to the north, and a greater extent of eolian deposits. It is also responsible for hypersaline conditions within Baffin Bay and Laguna Madre. The Rio Grande is the only large river in the region. Including endorheic basins that likely contributed to its sediment

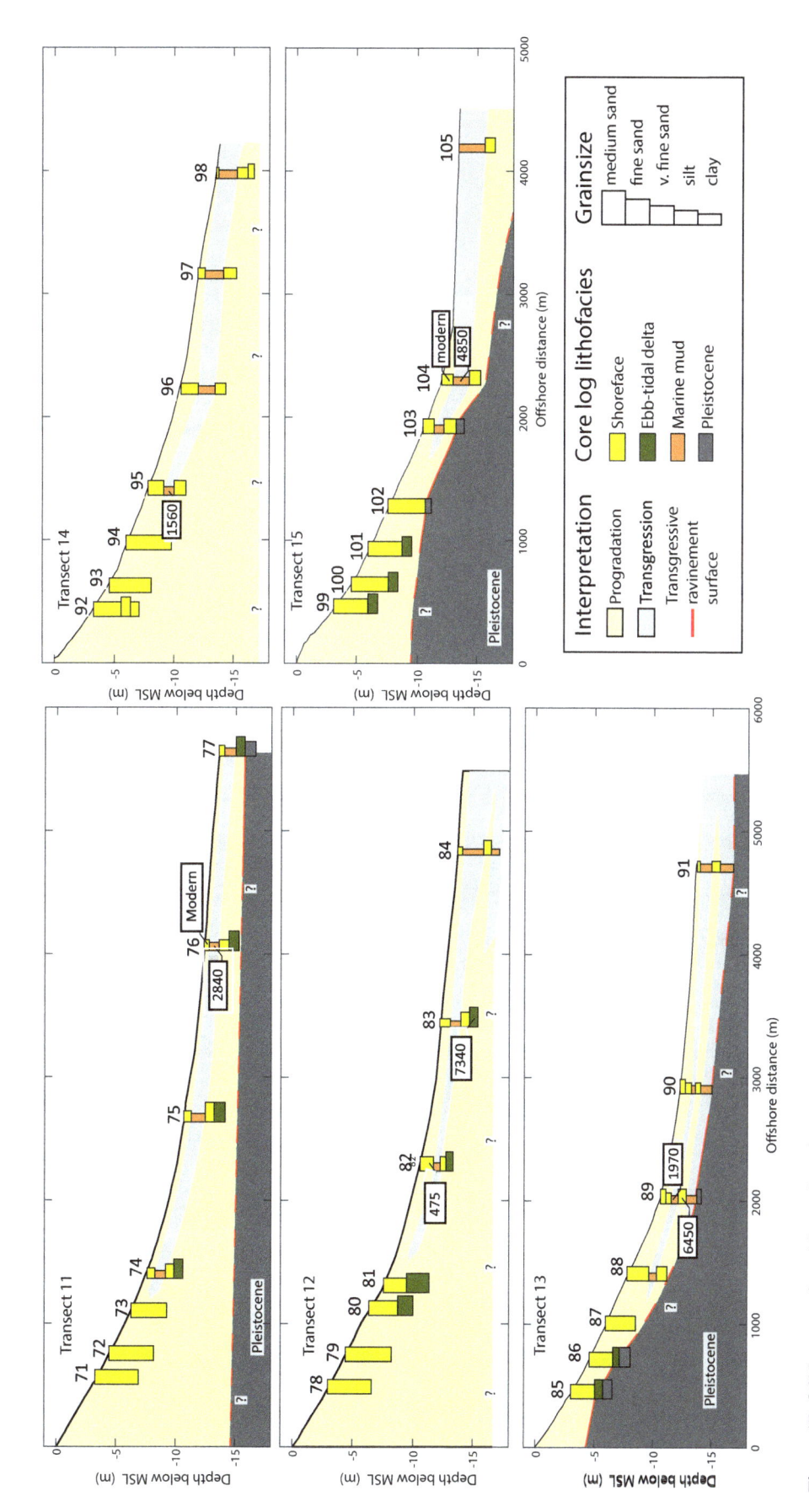

Figure 47. Offshore stratigraphic sections for core transects TR11–TR15 show two episodes of shoreface progradation and marine mud onlap separated by a phase of retrogradation and marine mud onlap (modified from Odezulu et al., 2020). Radiocarbon ages in cal yr B.P. are also shown. See Figure 45 for transect locations. MSL—mean sea level.

discharge when more humid conditions occurred, it has a drainage basin area of ~805,000 km^2, making it the second largest in the Gulf of Mexico region. Furthermore, its drainage basin is one with considerable relief and has experienced significant climate variability, which has resulted in considerable change in sediment discharge through time, manifest as shifts from large, fluvial-dominated deltas to sediment-starved coasts during the previous glacial-eustatic cycle (Banfield and Anderson, 2004; Fig. 6).

The Rio Grande Delta. During the MIS 2–1 transgression, the Rio Grande Delta stepped landward and formed a lobate, fluvial-dominated delta on the middle shelf (Banfield and Anderson, 2004; Fig. 28C). The age of this delta is not directly constrained by radiocarbon ages, but its location on the shelf and depth range indicates that it formed during the early Holocene between ca. 11,000 ka and ca. 9000 ka, or at about the same time that the Brazos and Colorado rivers formed fluvial-dominated deltas on the east Texas shelf (Anderson et al., 2016). It is likely that all three of these deltas were left isolated on the shelf during the ca. 9.6 ka MWP1c sea-level event (Fig. 4B). By mid-Holocene time, the Rio Grande Delta had shifted landward as a more shore-parallel, wave-dominated delta. The timing of this change is inferred to have been between ca. 7.0 ka and ca. 5.0 ka based on the water depths of this feature (Banfield and Anderson, 2004). This is similar to the Holocene Colorado River Delta's evolution. Morton and Winker (1979) and Banfield and Anderson (2004) argued that the landward retreat of these deltas resulted from decreasing water and sediment discharge in response to warmer and drier conditions at that time. Anderson et al. (2016) further suggested that the Rio Grande likely experienced large variations in sediment discharge to the Gulf of Mexico due to changes in the amount of sediment sequestered in the onshore alluvial valley and its numerous endorheic basins during the Holocene.

The modern Rio Grande Delta is a broad (~150-km-wide), lobate feature bounded on either side by large lagoons and multiple distributary channels (Fig. 48A). These channels are highly sinuous, with sandy point bars that extend to the current shoreline (Fig. 48B). Thus, the landscape is one that reflects significant wave reworking of the delta front that resulted in the formation of narrow barrier islands to the north and south.

One of the unique features of the south Texas inner continental shelf are bathymetric ridges that are oriented oblique to the current shoreline. Unlike sand ridge fields of the eastern Gulf, these features have more variable morphologies, with amplitudes of less than 2 m and irregular spacing. Cores collected from one of these features sampled a thin veneer (up to 115 cm) of shelly sand resting on delta mud, which indicates that these are not sand waves but rather reflect a paucity of sand on the inner shelf (Rodriguez et al., 2001).

Baffin Bay. Baffin Bay is unique among Texas bays in that it is not connected directly to the Gulf of Mexico via a tidal inlet (Fig. 49A). Freshwater drainage into the bay is from three creeks (Los Olmos Creek, San Fernando Creek, and Petronila Creek), with a combined drainage basin area of just 11,200 km^2. As a result of its relatively small and highly variable freshwater

Figure 48. Google Earth images show (A) Rio Grande Delta and (B) a close-up view of meanders and oxbow lakes of the Rio Grande near the coast. White box in panel A is location of image in panel B.

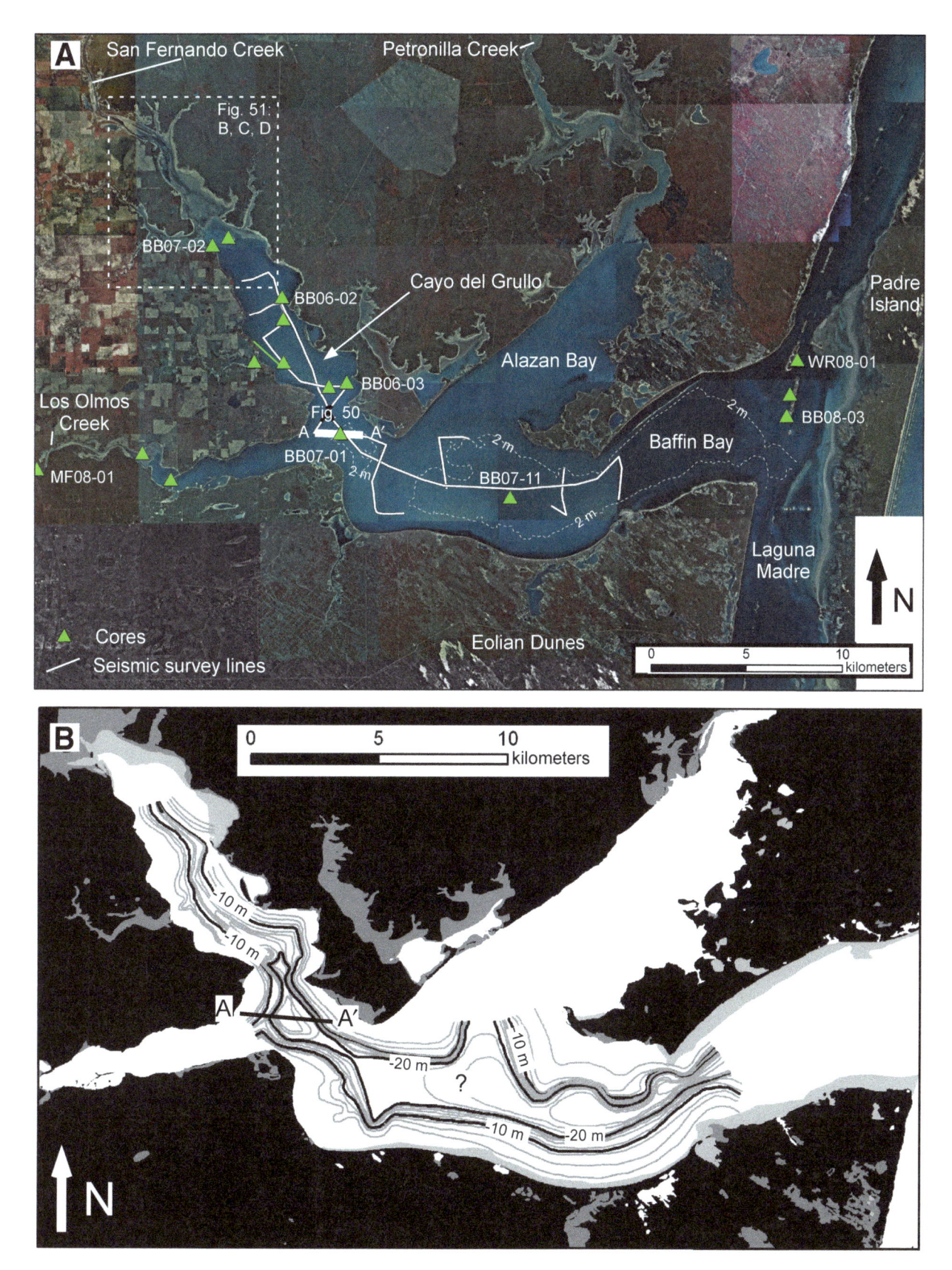

Figure 49. (A) Locations of seismic lines and sediment cores (green triangles) used to investigate Baffin Bay are shown. (B) Isopach map of Pleistocene surface beneath Baffin Bay. Both figures were modified from Simms et al. (2010).

input, and because of its location in a semi-arid setting, the bay experiences significant salinity variations between 2 ppt and 85 ppt and averages 40–50 ppt (Behrens, 1966). The result is a unique array of sedimentary facies, including evaporites, ooids, algal mats, and caliche deposits (Price, 1933; Behrens, 1963; Dalrymple, 1964; Driese et al., 2005). The thick caliche deposits provide resistant shorelines that have served to maintain a dendritic drainage pattern that extends to the bay (Behrens, 1963).

Behrens (1963) first acquired seismic data that were used to map the incised valley now occupied by Baffin Bay. His results were later augmented by Simms et al. (2010), who acquired additional seismic data to construct the map of the valley shown in Figure 49B. A seismic profile across the valley shows its u-shaped profile and indicates that it reaches depths of up to 30 m (Fig. 50), similar to other incised valleys of Texas with much larger fluvial discharge. This, plus the relatively thick fluvial section at the base of the valley, suggests that fluvial discharge and sediment supply to the valley during the late Pleistocene was greater than it is today.

Simms et al. (2010) conducted a detailed study of Baffin Bay using high-resolution seismic data and 18 sediment cores that sampled to depths of up to 17 m within the estuarine section (Fig. 51). They identified six seismic facies units and 15 sedimentary facies within the bay that together illustrate a flooding history similar to that of the other bays of the Texas Coast, including landward-stepping seismic and lithofacies associated with flooding surfaces. A total of 21 radiocarbon ages were used to examine the evolution of the bay.

Simms et al. (2010) observed that by ca. 8.0 ka, the valley had been flooded and an extensive bayhead delta was established (Fig. 51). At ca. 8.0 ka, a major flooding event resulted in expansion of the open bay setting and the bayhead delta shifting northward ~15 km. Given the limited age constraints on this flooding event, Simms et al. (2010) argued that it is likely the same ca. 8.2 ka flooding event impacted other bays of the Gulf Coast and bays globally (Rodriguez et al., 2010). Around 5.5 ka, a dramatic change in the Baffin Bay setting resulted in the character of the bay shifting from one similar to the other bays of the region to its current unique bay setting. This was also a time when several spits within the bay were flooded, suggesting a change in marine influence likely due to greater isolation of the bay in response to Padre Island shifting to a more landward location. This flooding event may correlate to the FS3 flooding event of Simms et al. (2008) in Corpus Christi Bay and to the transgressive event that is recorded in sediment cores offshore of Mustang and North Padre Islands, which began around 4.8 ka. Since that time, the depositional setting of Baffin Bay has not changed significantly, which indicates that the unique hypersaline setting has persisted (Simms et al., 2010). Livsey and Simms (2016) showed that, like the Nueces Delta to the north, Baffin Bay experienced climatically influenced changes during the middle to late Holocene. These changes included the backstepping of bay environments that appears to correlate with northerly shifts in the Intertropical Convergence Zone and positive phases of the Atlantic Multidecadal Oscillation (Livsey and Simms, 2016). Buzas-Stephens et al. (2014) analyzed down-core benthic foraminiferal assemblages in sediment cores from Baffin Bay and concluded that assemblage changes corresponded to climate cycles, specifically El Niño–Southern Oscillation and North American monsoon cycles. Livsey and Simms (2016) also quantified the amount of sand delivered to Baffin Bay through the late Holocene and found that bayhead delta and upper-bay mudflat backstepping events were associated with a decrease in the amount of sand delivered to the bay.

Lower Laguna Madre. Laguna Madre is the only hypersaline lagoon in the United States, stretching some 209 km from Corpus Christi Bay to the mouth of the Rio Grande (Fig. 1). The lagoon is divided by shallow flats, near its center, into what are commonly referred to as the upper and lower lagoon. It is mostly less than 1 m deep and connected to the Gulf of Mexico by two relatively small navigation channels, one at the southern end of the lagoon and the other at Port Mansfield near the center of the lower lagoon (Fig. 52A). The lower lagoon is ~32 km long and separated from the Gulf of Mexico by South Padre Island.

Wallace and Anderson (2010) acquired 37 vibracores along eight transects from the back-barrier and barrier shoreline westward to the center of lower Laguna Madre to examine the sedimentological record of storm impacts in the area (Fig. 52A). These cores sampled proximal washover sands resting on distal

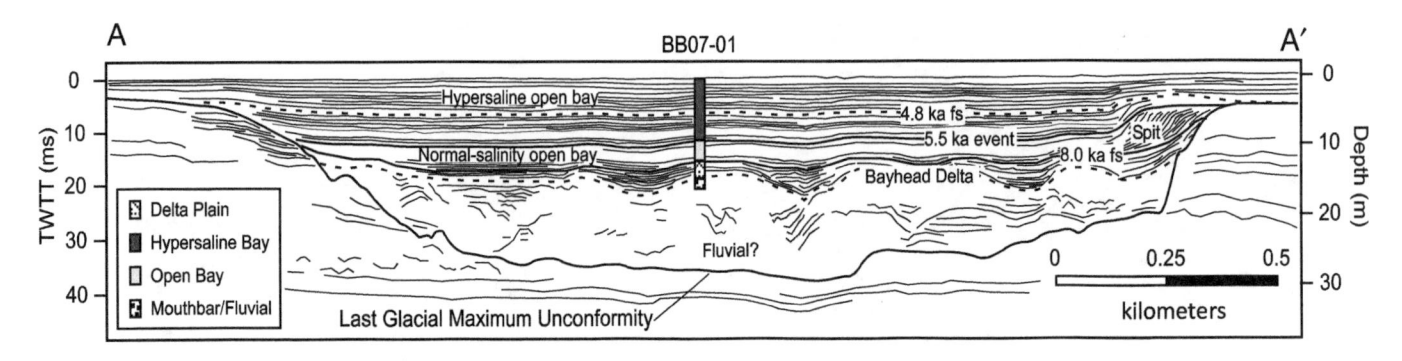

Figure 50. Interpreted seismic section (A–A′) shows seismic facies units and bounding flooding surfaces with ages (modified from Simms et al., 2010). See Figure 49 for profile location. fs—flooding surfaces.

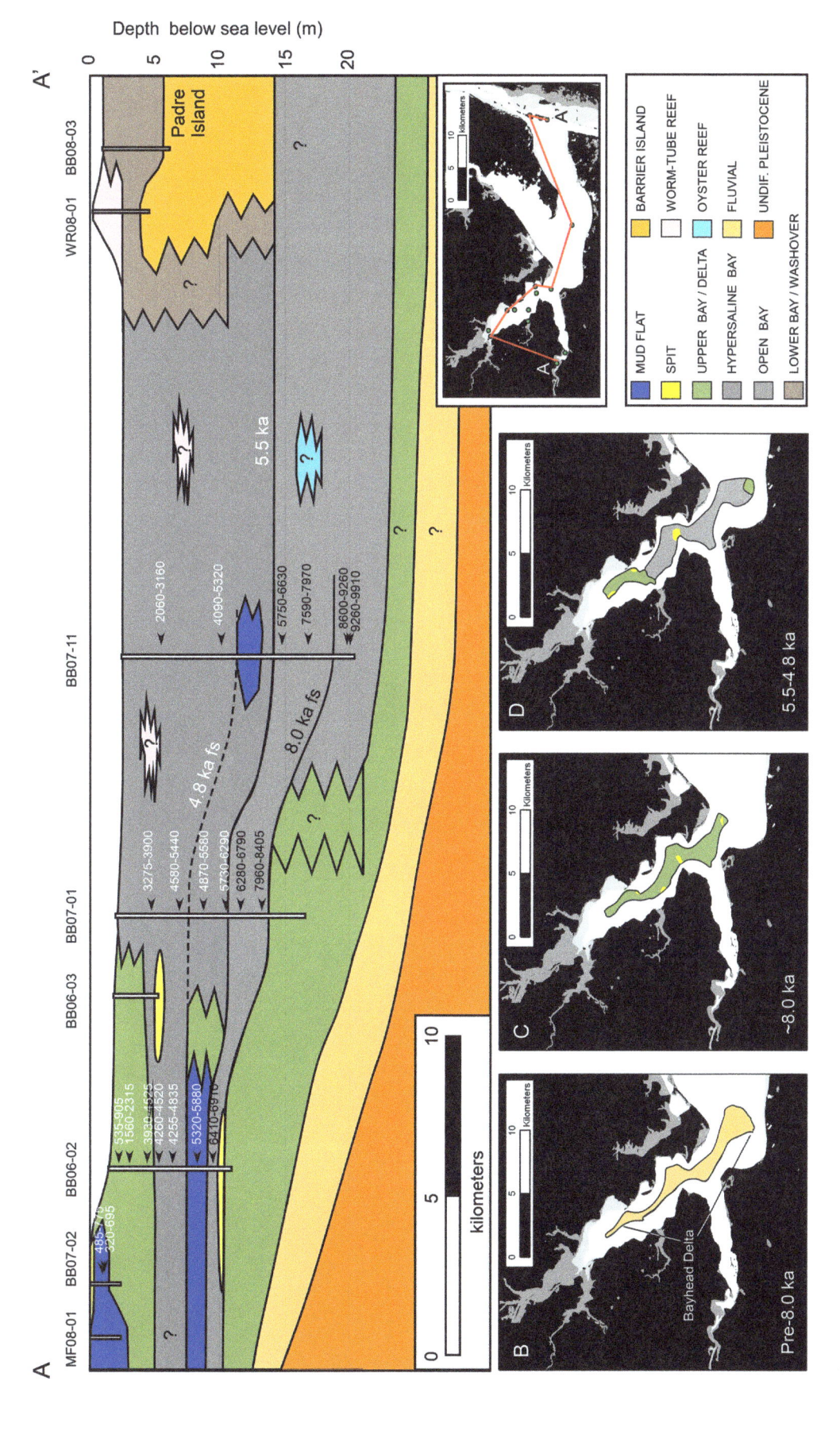

Figure 51. (A) East–west stratigraphic section for Baffin Bay was compiled from seismic data and drill cores, and the inset shows the profile location. Major seismic facies units are also shown along with bounding flooding surfaces and their radiocarbon ages. Paleogeographic maps illustrate stages of bay evolution for (B) pre-8.0 ka, (C) ca. 8.0 ka, and (D) 5.5–4.8 ka. Figure was modified from Simms et al. (2010).

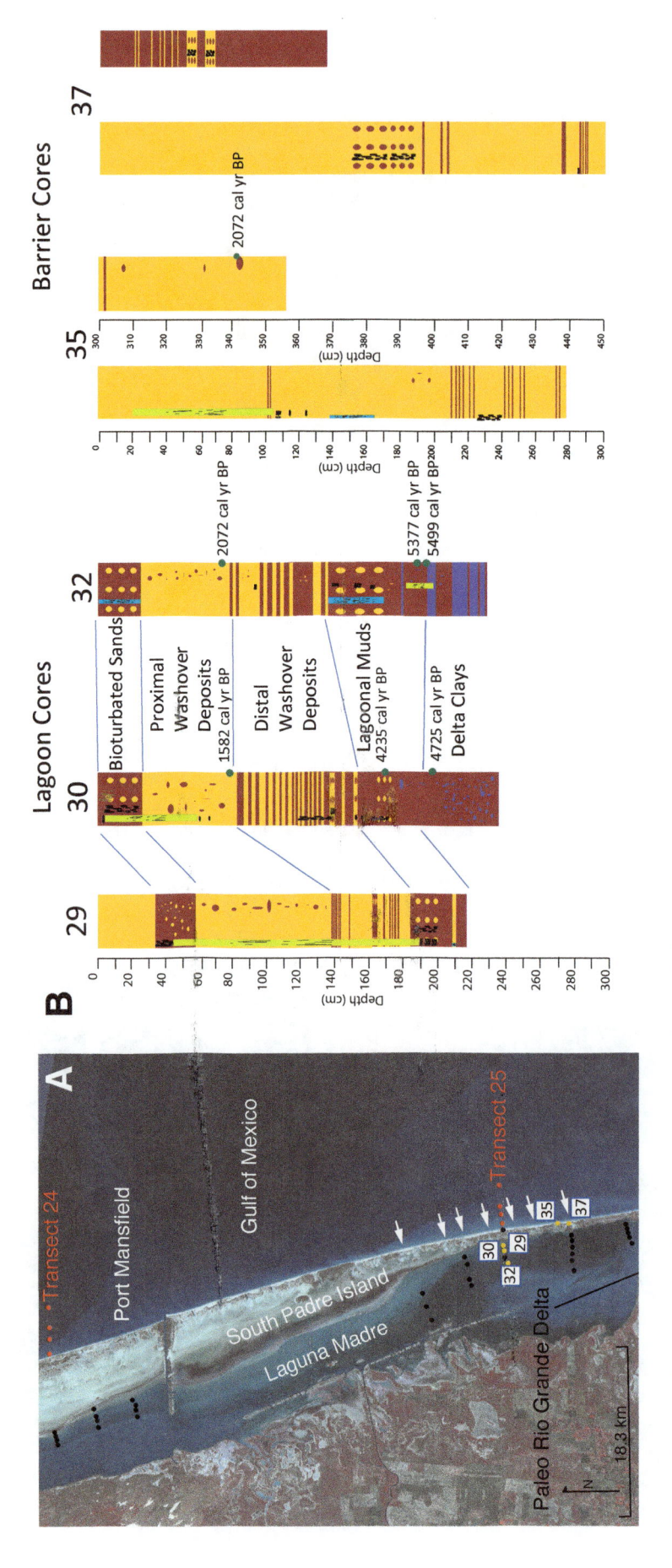

Figure 52. (A) Sediment core locations for Laguna Madre and South Padre Island are shown. (B) Lithological logs and environmental interpretations for representative cores from Laguna Madre and South Padre Island. Figure was modified from Wallace and Anderson (2010).

washover deposits that, in turn, rest on bay/lagoon mud and highly oxidized floodplain clays (Fig. 52B). Radiocarbon ages from four cores indicate that bay/lagoon muds were accumulating prior to ca. 5.5 ka and that the accumulation of distal washover deposits began shortly thereafter. Beginning at ca. 2.0 ka, there was a shift to proximal washover deposition in some cores, indicating that South Padre Island was located near its current location for at least the past two millennia.

South Padre Island. South Padre is a narrow, low barrier that is dominated by washover features. Two sediment cores through the island (cores PI-35 and PI-37, Fig. 52) sampled ~3.5 m of barrier sand, with distal washover resting in sharp contact on bay muds and highly oxidized red clay, which indicates a transgressive history for the island. A single sample from 340 cm depth in core PI-35 yielded a radiocarbon age of 2072 cal yr B.P. Distal washover deposits in core PI-30 range in age from 4235 cal yr B.P. to 1582 cal yr B.P., and in core PI-32 from 5377 cal yr B.P. to 2072 cal yr B.P. Another age constraint comes from the bayside of the island near Baffin Bay, where core NLM7 sampled proximal washover deposits that yielded an age of 2242 cal yr B.P. These ages are consistent with those from Laguna Madre proximal washover deposits that indicate the barrier was near its current location by ca. 2.0 ka.

Two offshore core transects were collected north and south of Port Mansfield (Fig. 53). Both transects are characterized by relatively steep shoreface profiles, and sediment cores ST 4 and ST 8 from the upper shoreface sampled thin (less than 1.6 m) sand with abundant shell material and shell hash resting in sharp contact on red clay. The remaining cores in these transects sampled thin (< 25 cm) sand units with abundant shell material also resting in sharp contact on red clay (Fig. 53). Due to the reworked nature of shoreface deposits, no radiocarbon ages were acquired from these offshore cores. However, the steep shoreface profile and thin and restricted extent of upper shoreface sand is consistent with a transgressive barrier as at Bolivar Peninsula, Follets Island, and Matagorda Peninsula.

South Padre Island currently has one of the fastest retreating coastlines in Texas, with an average annual rate of −2.5 m/yr (Paine et al., 2021). Based on historical rates of shoreline change, the volume of sand eroded from South Padre is estimated to be 312,800 m^3/yr, of which a significant amount is moved landward through overwash. Offshore and alongshore fluxes are minimal. This is significantly larger than the long-term overwash flux estimated from sediment cores collected within Laguna Madre (Wallace and Anderson, 2010), which suggests a modern acceleration of coastal retreat by washover as South Padre Island retreats landward.

The modern Rio Grande is currently delivering minimal sand to the coast, and its delta is experiencing landward retreat. This is in large part due to the construction of Falcon Dam in 1953 and another large dam on the river's largest tributary, the Rio San Juan, ~20 km north of where it merges with the Rio Grande. This, plus excessive surface and subsurface water usage by Texas and Mexico, have significantly reduced the flow of the river to the gulf (Benke and Cushing, 2005; Ewing and Gonzalez, 2016) and associated sand delivery to the coast. This, in turn, has significantly reduced sand supply to South Padre Island carried by prevailing northward-directed currents (Moore et al., 2021). Offshore cores have sampled mostly delta silts and clays (Fig. 53), so relatively little sand is being supplied to the coast through transgressive erosion of offshore sources. A recent study by Caudle et al. (2019) suggests that the Rio Grande Delta may have been experiencing slower retreat over the past two decades, resulting in an increase in beach and dune volume on South Padre Island. They attributed this increase in subaerial sand storage to a lack of tropical cyclone impacts to the area and sediment contribution from beach nourishment projects.

In summary, the south Texas Coast is a unique setting where climate has played a major role in coastal evolution. The evolution of Baffin Bay was initially dominated by sea-level rise and includes a record of punctuated rise during the early and middle Holocene. By late Holocene time, the bay shifted to a hypersaline estuary in response to increased aridity in the region. Padre Island is a transgressive barrier whose evolution has strongly influenced Laguna Madre and Baffin Bay. Today, this is one of the more rapidly changing areas of the LaTex Coast and the area that is impacted most by ongoing climate change.

DISCUSSION

Coastal Response to Climate Change and Decreasing Sea-Level Rise

The early Holocene record of coastal evolution in the northwestern Gulf of Mexico is fragmented due to transgressive erosion of offshore deposits and burial by marine and deltaic mud. This erosion results in a prominent transgressive erosion surface known as the TRS, which in Texas occurs at a water depth of −8 m to −12 m (Wallace et al., 2010). Below this depth, marine mud typically rests on Pleistocene deposits, and coastal deposits are absent. Exceptions to this are Sabine, Heald, and Shepard banks (Fig. 1), which are interpreted as drowned coastal barriers (Rodriguez et al., 1999, 2000). These barriers formed during periods of shoreline stability that lasted between 1000 and 2000 years and were followed by landward shifts of the shoreline of many kilometers. Other exceptions include the early Holocene Brazos, Colorado, and Rio Grande deltas and fluvial and estuarine strata preserved within incised valleys. The early Holocene Brazos, Colorado, and Rio Grande deltas were fluvial-dominated deltas that switched to wave-dominated at ca. 9.5 ka. These deltas were eventually abandoned as more of the sediment from the rivers was deposited in their onshore valleys. In the case of the Brazos and Colorado rivers, this occurred as sea level rose and sediments began to fill the broader valley margins and the upper portions of these valleys (Fig. 30). The early Holocene was also a time of transition from cool/wet to warm/dry conditions (Fig. 54), which may have influenced sediment discharge of these rivers.

The early Holocene sediment fill of the offshore Trinity River and Sabine River valleys is composed of correlative tidal inlet/

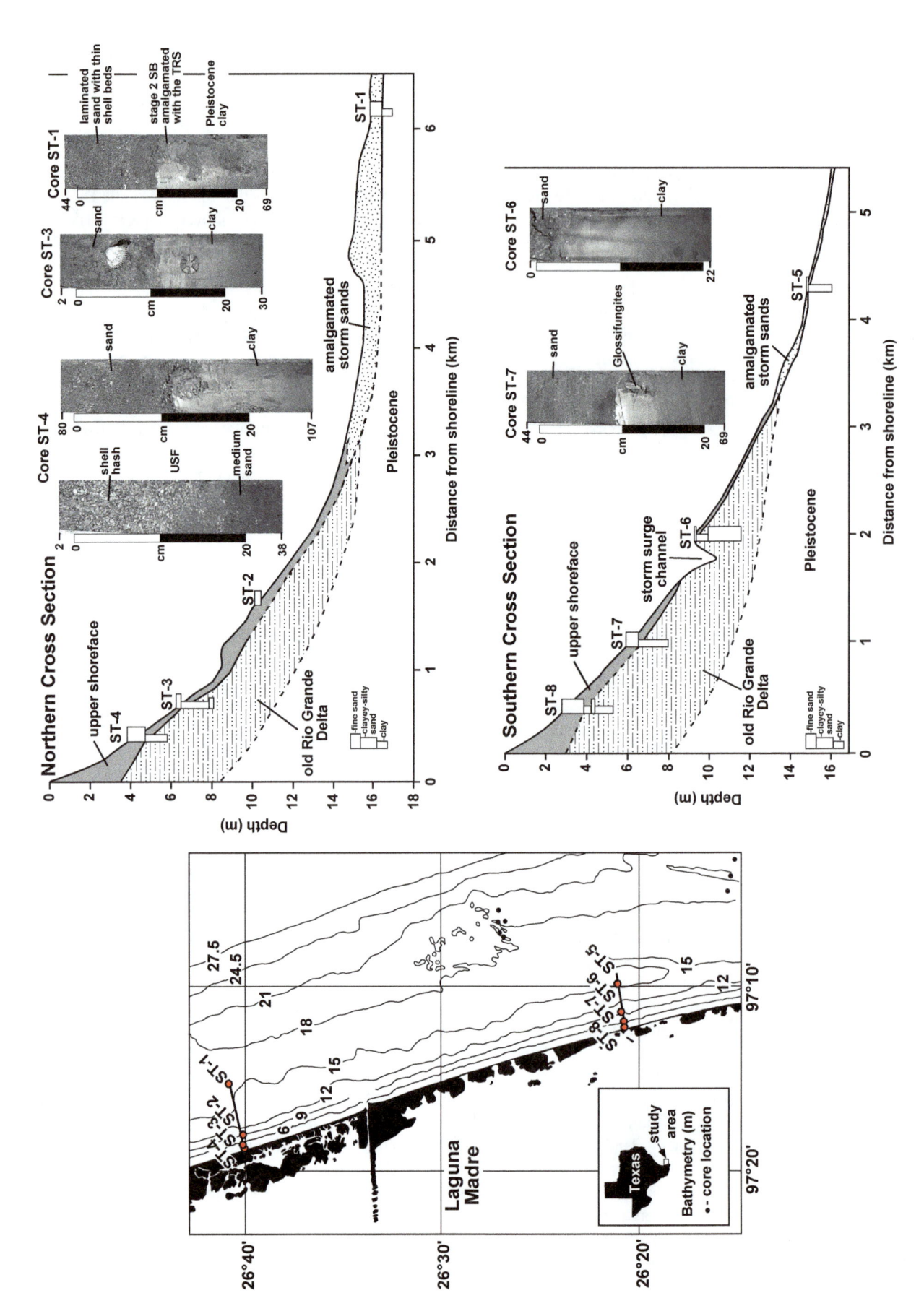

Figure 53. (Left image) Location map for core transects offshore of South Padre Island. (Right images) Stratigraphic sections derived from offshore core transects 24 (Northern Cross Section) and 25 (Southern Cross Section) (locations are shown in Figure 52). Figure was modified from Rodriguez et al. (2001). SB—sequence boundary; TRS—transgressive ravinement surface; USF—upper shoreface.

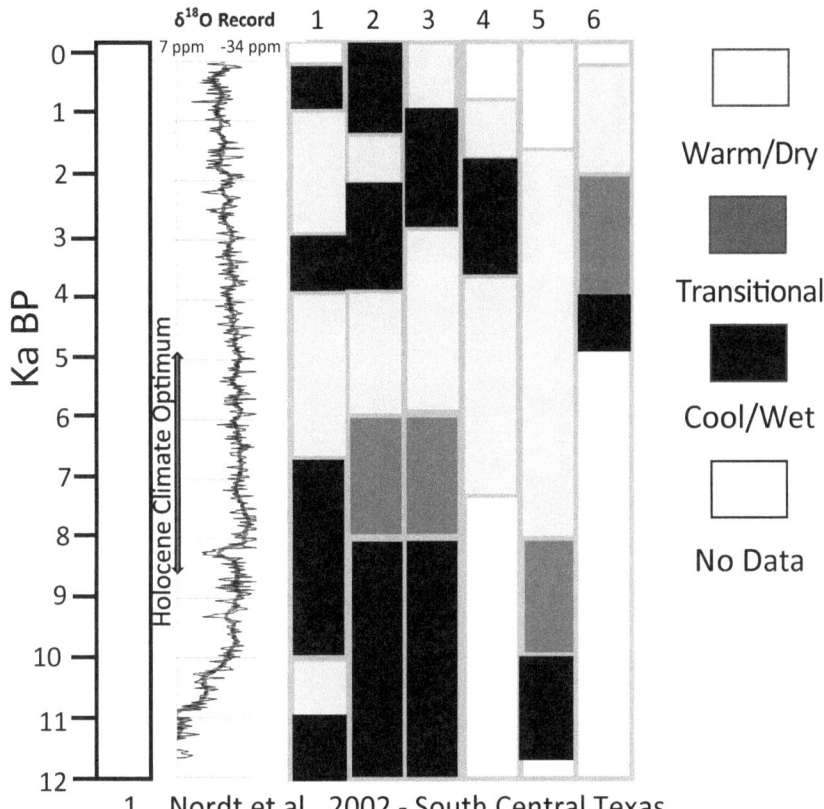

Figure 54. Holocene oxygen isotope record (data from Vinther et al., 2006, and Rasmussen et al., 2006) and paleoclimate summary chart for Texas are shown with sources listed.

1. Nordt et al., 2002 - South Central Texas
2. Nordt et al., 1994 - Central Texas
3. Toomey et al., 1993 - Central Texas
4. Humphrey and Ferring, 1994 - North Central Texas
5. Ferguson et al., 2018a - Corpus Christi Bay
6. Ferguson et al., 2018b - Trinity Bay

delta, middle bay, and tidal delta facies bounded by transgressive surfaces. Since multiple coastal environments were impacted by these flooding events, sea-level rise was the likely culprit. This landward-stepping stratigraphic architecture indicates landward shifts in the gulf shoreline and bayline of a few tens of kilometers (Fig. 11). Based on the depths of these flooding surfaces, they are interpreted as being late Pleistocene–early Holocene in age, but their exact ages remain uncertain (Thomas and Anderson, 1994). However, results from subsequent studies (Milliken et al., 2008b, 2008c; Anderson et al., 2008; Maddox et al., 2008; Simms et al., 2008, 2010; Troiani et al., 2011) of modern bays indicate widespread flooding events at ca. 9.8 ka to ca. 9.2 ka, ca. 8.9 ka to ca. 8.5 ka, ca. 8.4 ka to ca. 8.0 ka, and ca. 7.9 ka to ca. 7.5 ka, when bayhead deltas, open bays, and tidal deltas stepped landward, followed by periods when these environments stabilized and aggraded (Fig. 55). Correlation of flooding events among multiple bays supports previous arguments that punctuated sea-level rise was responsible for these flooding events. The magnitudes of these flooding events relative to rates of sea-level rise is consistent with numerical modeling results (Moran et al., 2017).

The 9.8–9.2 ka flooding event resulted in initial flooding of Galveston, Matagorda, Copano, Corpus Christi, and Baffin bays (Fig. 55). It is not recorded in all of the bays we studied due to the shallower depths of some incised valleys. At about the same time the offshore Colorado Delta, and possibly the Rio Grande and Brazos deltas, shifted landward and became wave-dominated deltas, the east Texas shoreline shifted from the location of Shepard Bank to Heald Bank, and the central Texas shoreline shifted to near its current location (Simms et al., 2006). This is believed to have been a global event (MWP-1c) and is recorded in the composite sea-level curve for the western Gulf of Mexico as potentially having been as much as 6 m of rise within two to three centuries (Fig. 4B).

During the 8.9–8.5 ka flooding event, the onshore Sabine and Calcasieu valleys were initially flooded, and bayhead deltas within Galveston Bay experienced an ~10 km landward shift

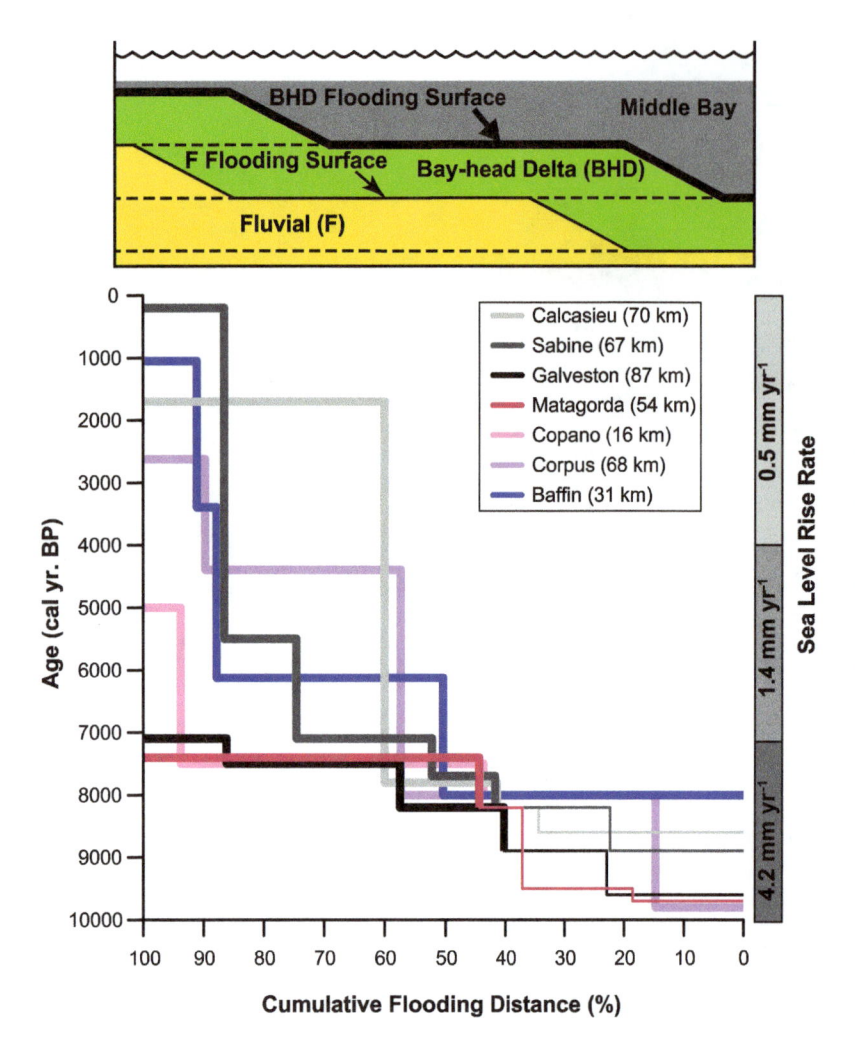

Figure 55. Summary chart of Holocene flooding events preserved in incised valleys is shown. Flooding surfaces recognized as bay-head delta overlain by middle bay facies are depicted with wide lines, and fluvial surfaces overlain by bay-head delta facies are depicted with thin lines (see conceptual model, top). Flooding distance is expressed as a percent of the total flooding distance (km) and presented parenthetically after the site name in the legend. Average rates of Holocene sea-level rise are after Milliken et al. (2008a) and exclude events such as the rapid rate of rise at 8200 cal yr B.P.

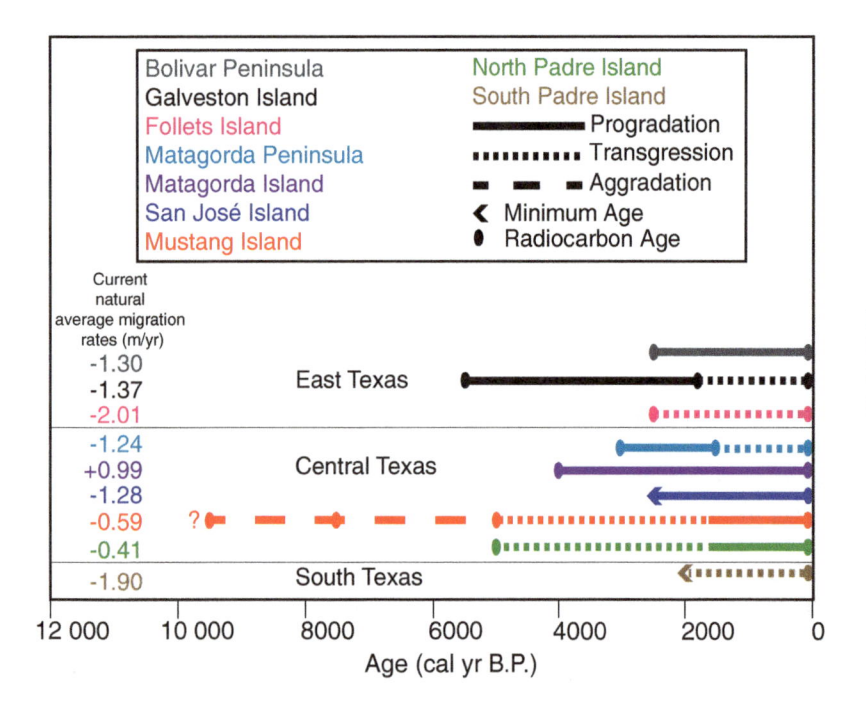

Figure 56. Coastal barrier evolution of the Texas Coast is summarized (modified from Anderson et al., 2014). Also shown are the current natural average migration rates (from Paine et al., 2021).

(Fig. 55). This was also around the time of a twofold increase in valley aggradation within the Brazos and Colorado valleys, marking another onshore shift in sedimentation (Taha and Anderson, 2008).

The 8.4–8.0 event is the best documented of the early Holocene flooding events (Fig. 55). It is recorded in every bay studied from Lake Calcasieu to Baffin Bay as a 4–15 km landward shift of bayhead deltas (Rodriguez et al., 2010) and on the east Texas continental shelf by the drowning of Heald Bank. In Corpus Christi Bay, the bayhead delta shifted more than 20 km landward, and there was widespread destruction of oyster reefs in the bay (Simms et al., 2008; Goff et al., 2016; Rice et al., 2020). A micropaleontological analysis of a sediment core from upper Corpus Christi Bay showed a nearly fivefold increase in dinoflagellate cysts associated with this flooding surface, which indicates that salinities in the bay abruptly increased to open marine levels (Ferguson et al., 2018a). These results further indicate that Mustang Island was not an effective salinity barrier during this event and that it took between two and three centuries for the bay to return to normal estuarine conditions.

The final early Holocene flooding event occurred between 7.9 ka and 7.5 ka (Fig. 55) and coincides with a global sea-level event at 7.6 ka (Yu et al., 2007). In Galveston Bay, the Trinity bayhead delta shifted landward ~25 km, which equates to a rate of landward retreat of 6.3 km/century (Fig. 13). Matagorda and Copano Bays were also significantly impacted by this event (Fig. 34).

The middle Holocene was marked by a three-fold decrease in the rate of sea-level rise, from 4.2 mm/yr to 1.4 mm/yr (Fig. 4B). This resulted in greater regional variability in coastal change as other factors, such as riverine sediment supply and dispersal, offshore mud blanket development, and antecedent topography, began to dominate. By ca. 7.0 ka, sea level had risen to approximately −7 m, and the shoreline of central Texas was near its current location in contrast to the flatter western Louisiana and east Texas region, where the shoreline was located near Sabine Bank. By ca. 5.3 ka, an embayment developed on the east Texas Coast separating the shoreline at Sabine Bank from the newly formed Galveston Island shoreline (Fig. 1). The Brazos and Colorado River mouths shifted landward to form the current delta headland. A significant portion of the sediments transported by these rivers was now being deposited within their onshore valleys, where aggradation rates kept pace with sea-level rise to maintain fluvial outlets at the coast (Taha and Anderson, 2008).

The mid Holocene was also a period of significant climate change in Texas and was marked by a transition from cool and wet to warm and dry conditions (Fig. 54). This had a significant impact on south Texas bays. Between ca. 5.8 ka and 5.4 ka, Baffin Bay experienced a dramatic change to its current hypersaline conditions (Simms et al., 2010). Pollen records from core CCB02-01, collected in the upper part of Corpus Christi Bay, show a gradual increase in arboreal pollen around this time, indicating more arid conditions (Ferguson et al., 2018b). To the north, Troiani et al. (2011), who used grain-size changes in sediment cores from Copano Bay as a paleo-wind indicator, observed windier conditions from 5.2 ka to 4.1 ka, which they attribute to a drier climate for the region.

In central Texas, the mid-Holocene shoreline was located near its current location and coastal progradation was occurring between Matagorda Peninsula and North Padre Island. This phase of barrier growth was likely nourished by sand eroded from the offshore Colorado and Rio Grande deltas (Odezulu et al., 2020). It ended with a transgressive phase of barrier evolution that began at ca. 4.9 ka at Mustang Island.

By late Holocene time, the average rate of sea-level rise in the western Gulf of Mexico had decreased to 0.5 mm/yr (Fig. 4B). Modern bays had begun to take on their current shape as the broader shoulders of the valleys they occupy were gradually flooded (Figs. 13, 15, 17, 34, and 37). There were no widespread flooding events that impacted bays across the entire region. In Galveston Bay, the Trinity bayhead delta experienced ~4 km of progradation since ca. 3 ka, which indicates that sediment supply was large enough to outpace slow sea-level rise. During the same time the Trinity bayhead delta was prograding, Sabine Lake and Calcasieu Lake experienced major flooding episodes between ca. 2.5 ka and 1.8 ka, when their bayhead deltas shifted northward ~15 km (Milliken et al., 2008b, 2008c).

Baffin Bay and Corpus Christi Bay experienced multiple flooding events during the middle to late Holocene. These events occurred at ca. 5.0 ka and ca. 3.2 ka in Corpus Christi/Nueces Bay (Rice et al., 2020) and 6.5–5.7 ka, 5.2–4.9 ka, 3.0–3.8 ka, 2.7–2.1 ka, and 1.1–1.0 ka in Baffin Bay (Buzas-Stephens et al., 2014; Livsey et al., 2016) and are attributed to diminished sediment supply tied to climate changes recorded by independent local climate proxies. The bayhead delta and upper-bay mudflats of Nueces and Baffin bays, respectively, also experienced progradation during intervening time periods including the period leading up to modern times.

There is evidence for increased sediment discharge from the Brazos, Colorado, and Rio Grande rivers to the Gulf of Mexico during the late Holocene in the form of major expansion of the Texas Mud Blanket (Weight et al., 2011). This increase was likely associated with increased climate variability in the form of higher frequency alternations between cool/wet and warm/dry conditions across the western half of the state (Fig. 54). These climate oscillations were also recorded in the more southerly and westerly estuaries of Baffin and Corpus Christi Bays.

The late Holocene was also a time of variable evolution of coastal barriers and chenier plains, as summarized in Figure 56. Especially notable is the fact that Galveston Island, Follets Island, and Matagorda Peninsula all experienced transgression over the past millennium while Matagorda Island, San Jose Island, Mustang Island, and the northern portion of North Padre Island experienced progradation. This progradation is believed to have resulted, at least in part, from a westward shift in the Colorado River mouth during this time and from shoreface erosion of these relatively thick barriers.

The chenier plains have formed since ca. 3.0 ka, after the Calcasieu and Sabine valleys were filled to within a few meters of sea level (Milliken et al., 2008b, 2008c). The age and frequency of ridge formation of the Sabine Chenier Plain was ~85 years, which suggests that storms played a key role in their formation and that sand for ridge formation was readily available from nearby fluvial channels and tidal deposits. Widespread erosion of Bolivar Peninsula occurred between ca. 0.8 ka and ca. 0.6 ka, presumably due to one or more severe storm impacts. This was followed by a rapid phase of barrier healing, which is when most of the modern beach ridges of the peninsula were formed. The frequency of ridge formation during this healing phase was ~80 years, consistent with the rate of the Sabine Chenier Plain.

Controls on Coastal Evolution

Spatial and temporal differences in coastal evolution along the LaTex Coast indicate that multiple factors have influenced coastal evolution. These include the following:
1. Antecedent topography
2. Sea-level rise
3. Sediment supply and dispersal
4. Oceanographic influences and the Texas Mud Blanket
5. Impacts from severe storms
6. Anthropogenic impacts

Antecedent Topography

Our results demonstrate that coastal evolution has been strongly influenced by antecedent topography that resulted from MIS 5–2 delta development and fluvial incision of the continental shelf. In particular, the evolution of bays was strongly regulated by the antecedent topography of the incised valleys they occupy (Rodriguez et al., 2005; Simms and Rodriguez, 2014, 2015). The evolution of coastal barriers was also influenced by relief on the Pleistocene surface and tidal erosion surfaces as well as the continental shelf gradient (Anderson et al., 2014; Otvos, 2018).

Incised Valleys. The western Louisiana–east Texas continental shelf is relatively wide with a gentle profile (Fig. 1). This physiography is largely due to the presence of large, late Pleistocene deltas and fluvial floodplains that prograded across the shelf as sea level fell during MIS 5–2 (Fig. 6). In contrast, no deltas formed on the central Texas continental shelf, resulting in an embayment with a relatively steep shelf gradient (Fig. 1). The relief on the continental shelf was concurrently and subsequently modified by fluvial incision, a response to MIS5–2 sea-level fall and subaerial exposure. The result was a branching network of channels on the western Louisiana shelf, remnants of late Pleistocene fluvial-deltaic drainage in the region, in contrast to the widely spaced, incised valleys of the ancestral Calcasieu, Sabine/Neches, Trinity, Brazos, and Colorado rivers on the east Texas continental shelf (Fig. 9). The central Texas continental shelf is, in turn, characterized by fluvial valleys that are traceable only to the middle shelf (~60 m water depth), where there is an abrupt step in the Pleistocene surface. Seaward of this location, valley morphology is masked by the overlying Texas Mud Blanket, but the valleys appear to converge into a single wide, low-relief valley (Berryhill, 1987; Eckles et al., 2004). The MIS 2 fluvial relief on the south Texas shelf is dominated by a single valley cut by the ancestral Rio Grande (Berryhill, 1987; Banfield and Anderson, 2004).

The Calcasieu and Sabine rivers both formed relatively narrow, seaward-converging valleys that exhibit complex Holocene flooding histories (Figs. 14B–16B). To the west, incised valleys are characterized by deep, narrow incisions bounded by broad, shallow flanks that typically have terraces (Figs. 12B, 33, and 36B). The exception is Baffin Bay, which occupies a deep, narrow valley (Fig. 49B). During the early Holocene, sedimentation was confined to the deep, narrow portions of these valleys. This resulted in the formation of thick, elongate bayhead deltas that stepped landward in response to punctuated sea-level rise (Fig. 11). By middle and late Holocene time, rates of landward bayhead delta migration decreased as the broader flanks of the valleys were flooded. Valley shape, in particular the extent and elevation of fluvial terraces, began to play a greater role in bay evolution (Rodriguez et al., 2005). The change from narrow, deep bays to broad, shallow bays led to diminished tidal influence recorded by smaller flood tidal deltas and more widespread middle bay mud accumulation. The formation of coastal barriers and associated tidal inlets contributed to these changes.

Coastal Barriers. The modern coastal barriers of Texas formed at different times and have had different growth and retreat patterns (Fig. 56). This variability was in large part a result of differences in shelf gradient, which controlled rates of landward migration, and the antecedent topography on which these barriers formed. Recent results from numerical modeling studies coupled with sediment core analyses support the important role of antecedent topography on barrier morphodynamic behavior (Moore et al., 2010; Lentz et al., 2013; Brenner et al., 2015; Raff et al., 2018; Shawler et al., 2020).

During the early–middle Holocene, the coast migrated in step-wise fashion across the relatively gentle western Louisiana and east Texas shelf, pausing periodically as the ancestral Shepard Bank, Heald Bank, and Sabine Bank coastal barriers formed. Over the same time period, Matagorda Island, San Jose Island, Mustang Island, and North Padre Island migrated more slowly landward across the relatively steep central Texas shelf. This, coupled with the relatively high sand supply from adjacent eroding deltas, contributed to the development of the thicker (10–35 m), more stable barriers of the central Texas Coast relative to the thinner, less stable barriers of the east Texas and south Texas coasts. With this said, there is considerable variability in barrier thickness along the coast, and this variability is largely due to the antecedent topography on which these barriers rest.

There are two important components of the antecedent topography of the LaTex Coast. The first is the result of fluvial erosion created during the MIS 5–2 sea-level fall and lowstand (Fig. 9). The incised valleys formed during this time were flooded to create bays that have remained stationary throughout the MIS 1 transgression. Most coastal barriers, however, rest on

Holocene bay, deltaic, or shoreface deposits, so barrier development did not occur until sediments had filled most of the valleys. Therefore, the relief at the base of coastal barriers has been further modified by tidal erosion.

Galveston Island and Bolivar Peninsula are thickest where they overlie a deep tidal incision surface associated with the Bolivar Roads tidal inlet (Fig. 20) that has remained in its current location throughout the formation of these barriers (Siringan and Anderson, 1993). In the case of Bolivar Peninsula, the eastern portion of the barrier rests on relatively shallow Pleistocene surface, but the barrier's thickness increases abruptly near its west end, where it occupies the Bolivar Roads tidal incision. Assuming a constant rate of sand supply and longshore sand transport, the rate of growth of the peninsula would have decreased markedly as sand began filling this deep incision. Construction of a large jetty at the west end of Bolivar Peninsula in the early 1900s effectively blocked sand delivery to the tidal inlet, resulting in seaward progradation of the far west end of the peninsula.

Galveston Island, like Bolivar Peninsula, is thickest where it fills a deep Bolivar Roads tidal erosion surface and decreases in thickness to the west (Fig. 20). The 8-Mile Road transect, which in part is constructed from drill cores that penetrated the island, shows that at this location the island rests on a seaward-sloping Pleistocene surface that mimics the current offshore profile, which is an erosional surface cut during transgression (Fig. 24). As the island prograded, sedimentation of lower shoreface muds filled much of the relief on the Pleistocene surface, and the rate of island progradation increased from an average rate of ~0.4 km/ka between 5.3 ka and 2.6 ka to ~1.25 km/ka after ca. 2.6 ka. Since the depth of the Pleistocene surface decreases toward the west, there is a decrease in sediment accommodation and barrier thickness in that direction. Associated with this change in barrier thickness is a westward decrease in barrier width and height and an increase in storm washover features.

Similar to east Texas, significant variations in barrier thickness along the central Texas Coast are attributed mainly to antecedent topography. Although the stratigraphic record of its early development has been eroded, Matagorda Peninsula is believed to have had a history similar to that of Bolivar Peninsula, as westward accretion of the peninsula across a deep tidal inlet resulted in a decrease in its rate of lateral accretion as more sand was deposited in the Pass Cavallo tidal inlet (Fig. 40). Matagorda Island, San Jose Island, and Mustang Island are relatively thick islands that are believed to be composed largely of tidal deposits. These tidal deposits rest on laterally extensive tidal erosion surfaces that formed as a result of prolonged shoreline stability and sustained inlet migration (Figs. 40, 43, and 45B). These differences in barrier thickness have been a major influence on their relative stability. For example, Matagorda Peninsula has experienced significant transgression during the late Holocene while Matagorda Island has experienced growth and stability.

Mustang Island is another example of a thick barrier island composed largely of tidal deposits resting on an irregular tidal erosion surface (Fig. 45C). Unlike Matagorda Island, its thickness varies considerably along the coast, which reflects the lateral shifting of tidal inlets, as shown by radiocarbon ages from drill cores through the island (Simms et al., 2006).

Sea-Level Rise

Although the late Pleistocene and early Holocene record of coastal evolution in the western Gulf of Mexico is incomplete, there is a clear pattern of episodic growth of deltas, coastal barriers, and bay environments followed by rapid transgression or landward stepping. This style of coastal evolution calls for a forcing mechanism that impacts all of these environments contemporaneously, and punctuated sea-level rise is the most obvious control (Thomas and Anderson, 1994; Anderson et al., 2014; Mellett and Plater, 2018).

Punctuated sea-level rise during the late Pleistocene and early Holocene is well documented, both globally and in the western Gulf of Mexico (Fig. 4B). On the central Texas continental shelf, reef terraces record episodes of punctuated sea-level rise on the order of 5–6 m/ka during the late Pleistocene (Khanna et al., 2017), which is consistent with published sea-level curves (e.g., Fairbanks, 1989; Bard et al., 1990, 1996; Deschamps et al., 2012). Sea-level curves for the Gulf of Mexico suggest four episodes of punctuated sea-level rise for the early Holocene (Fig. 4B). The magnitude of these events was in the range of 2.0–4.0 m within one to two centuries, with rates on the order of 10–20 mm/yr (Milliken et al., 2008a; Donoghue, 2011; Törnqvist et al., 2020). Bayhead deltas were particularly sensitive to these episodes of rapid rise because of their extensive delta plains and modest sediment accumulation rates (Fig. 11). The punctuated sea-level rises also resulted in the demise of offshore fluvial-dominated deltas of the Brazos, Colorado, and Rio Grande rivers and landward-stepping barrier islands and tidal deltas.

Results from numerical modeling indicate that rates of rise of 10 mm/yr to 20 mm/yr in magnitude can cause significant increases in the rate of barrier island retreat, further leading to reductions in their width and height and renders them even more vulnerable to overwash (Moore et al., 2010; Lorenzo-Trueba and Ashton, 2014; Cowell and Kinsela, 2018; Murray and Moore, 2018; Ashton and Lorenzo-Trueba, 2018; Rodriguez et al., 2018). A case in point is Follets Island. Using generic island characteristics similar to those of Ashton and Lorenzo-Trueba (2018), Odezulu et al. (2018) determined that an increase in the rate of sea-level rise of 1–10 mm/yr would result in a four-fold increase in the magnitude of shoreline retreat for the island within 100 years. This implies a similar impact for Matagorda Peninsula and South Padre Island.

Results from numerical modeling and experimental studies have shown that periodic shoreline retreat can occur under constant sea-level rise due to time lags between the shoreface and overwash (Lorenzo-Trueba and Ashton, 2014), periodic overwash can lead to punctuated shoreline retreat (Rodgers and Paola, 2021), dune dynamics can drive discontinuous retreat (Reeves et al., 2021), and complex interactions between autogenic processes and transgressive response to rapid sea-level rise exist (Ciarletta et al., 2019). However, we observe a joint response of barriers, tidal inlets, and bay environments that indicates pulsed sea-level rise as the agent responsible for the observed landward-stepping stratigraphic record, as originally argued by Thomas and Anderson (1994). This interpretation is also consistent with results from Mellett and Plater (2018), who reviewed the literature on drowned late Pleistocene and Holocene barriers and found that the vast majority were formed during the early and middle Holocene, when the rate of sea-level rise was relatively high. It is also consistent with geomorphic evidence for large landward steps in grounding line features on the Antarctic continental shelf during the early Holocene (Bentley et al., 2014; Bart et al., 2017; Prothro et al., 2020).

The best documented Holocene punctuated sea-level event in the western Gulf occurred at ca. 8.2 ka. It resulted in significant landward retreat of bayhead deltas across the study area and globally (Rodriguez et al., 2010). This is approximately the time that Heald Bank was overstepped and isolated on the shelf. The magnitude of rise associated with this event is estimated to have been 0.4–1.2 m globally (Kendall et al., 2008) and between 0.2 and 0.56 m within the Gulf of Mexico (Li et al., 2012) and spanned a few centuries. Törnqvist and Hijma (2012) argue that rapid draining of glacial Lake Agassiz in north-central North America was a significant contributor to the 8.2 ka sea-level event.

During the middle Holocene, the average rate of sea-level rise in the western Gulf of Mexico was 1.4 mm/yr, and by the late Holocene the rate decreased to only 0.5 mm/yr (Fig. 4B). However, the scatter in data points in this curve prohibits reliable estimation of the magnitudes and rates of decadal to century-scale change. More recent results from a study of algal mats in Baffin Bay have resulted in sub-meter resolution (Livsey and Simms, 2013; Fig. 4B). These results revealed a possible punctuated event at ca. 5.0 ka but indicate a fairly steady rate of rise during the late Holocene. The scatter shown in the Livsey and Simms (2013) sea-level curve is approaching the range of the natural variability in sea level that results from meteorological and oceanographic influences.

Historical rates of sea-level rise in the Gulf of Mexico (Kolker et al., 2011), as determined from tide-gauge records and satellite data, are near the global average rate of 3.0 mm/yr (e.g., Chen et al., 2017; Oppenheimer et al., 2019), which is fast approaching the early Holocene rate (Fig. 4B). Compelling evidence for historical acceleration of sea-level rise also comes from research aimed at constructing accurate sea-level curves for the past few millennia (Englehardt et al., 2011; Kemp et al., 2011). Brain et al.'s (2015) sea-level curve from the North Carolina coast and compilation of other North Atlantic records shows that the most rapid change in the rate of sea level in the past ~2 k.y. occurred in the past two centuries. These results are consistent with estimates of the contributions to sea-level rise due to warming and expansion of the oceans and contributions from the melting of existent glaciers and ice sheets (Cazenave et al., 2009). The Greenland and West Antarctic ice sheets are currently experiencing rapid changes that could re-introduce punctuated sea-level events similar to those that occurred prior to and during the early Holocene (Siegert et al., 2020). Thus, sea-level rise is still considered the main threat to the world's coast.

Sediment Supply and Dispersal

Offshore Sediment Supply. Offshore sediment core transects from all along the Texas Coast have shown that the toe of the shoreface, which occurs at a water depth between −8 m and −10 m, coincides with the depth at which marine mud onlaps shoreface deposits, marking an erosion surface associated with storm wave base (Siringan and Anderson, 1994; Rodriguez et al., 2001; Wallace et al., 2010). During transgression this wave-cut surface migrates landward to form a widespread, relatively planar unconformity known as the transgressive ravinement surface (TRS). Erosion of late Holocene and modern deltas of the Brazos and Colorado rivers attests to the efficiency of transgressive ravinement (Fig. 31). The recycling of sands from these offshore deltas and fluvial channels during transgression has been the dominant source of sand to the LaTex Coast. Anderson et al. (2016) estimated that as much as 36.6 km^3 of sand could have been derived from erosion of these Pleistocene deposits, compared to ~13 ± 3 km^3 of sand that makes up the modern coastal cheniers, peninsulas, and barrier islands of the LaTex region. Some of the remaining sand has been sequestered in fluvial valleys, tidal deltas, and shoreface deposits.

The central Texas Coast highlights the importance of offshore deltas as sediment sources. During the early Holocene, a deep embayment in the shoreline between the Colorado and Rio Grande deltas existed (Fig. 28), and erosion of these deltas yielded large quantities of sand for coastal nourishment (Odezulu et al., 2020). By the middle Holocene, the rate of sea-level rise had decreased, and sand that was sourced from these deltas nourished a phase of pronounced coastal progradation that ended during the late Holocene when these offshore sand bodies were buried by the TMB.

Transgressive barriers, including Bolivar Peninsula, Follets Island, Matagorda Peninsula, and Padre Island, are composed of thin sand bodies (less than 2.5 m thick) that rest directly on Holocene back-barrier and Pleistocene deposits. This indicates that landward movement of these shorelines has outpaced the rate of translation of the TRS, resulting in this shallow erosion surface. This is because rates of shoreline translation and TRS translation are controlled by different processes. Shoreline migration over decadal to centennial timescales is largely controlled by the relative rate of sea-level rise versus sediment supply and by fair-weather wave climate. Erosion at the toe of the shoreface and TRS formation occurs at storm wave base and is therefore controlled by the rate of sea-level rise and storm frequency and magnitude with landward migration of this surface occurring over centennial to millennial timescales.

In the case of transgressive barriers that migrate landward across relatively shallow transgressive surfaces, less sand is recycled from the shoreface back into the coastal barrier. We suggest that the shallow erosion surface associated with Bolivar Peninsula, Follets Island, Matagorda Peninsula, and Padre Island indicates a deficit in sand supply necessary to maintain these barriers and an associated decrease in island elevation and increase in overwash. Thus, the presence of a shallow transgressive surface is an indication that a barrier is in the final stages of its life cycle, barring a decrease in the rate of sea-level rise or an increase in sand supply. How long the barrier will exist is largely dependent on its overwash rate, which is controlled mainly by back-barrier accommodation within bays and lagoons and barrier height.

By late Holocene time, offshore deltas were buried in marine mud, resulting in a decrease in sand supply to the coast (Odezulu et al., 2020). Subsequently, sand supply to the coast has been reduced to direct input from the Brazos, Colorado, and Rio Grande rivers, reworking of fluvial channels by transgressive ravinement, and reworking of coastal barriers. Of these, direct sand delivery from rivers has been significantly reduced in historical time due to anthropogenic alteration of these drainage systems.

An extensive network of fluvial channels occurs on the inner shelf of western Louisiana (Fig. 9), and seismic records show that these channels were eroded during the late Holocene (Suter and Berryhill, 1985; Milliken et al., 2008b). Thus, these channels are believed to have been an important source of sand for the beaches and chenier plains of western Louisiana and east Texas (McBride et al., 2007; Anderson et al., 2014). Elsewhere on the continental shelf, the spacing of fluvial channels on the inner shelf is greater, and the upper portions of most channels are composed of marine mud and fine-grained estuarine sediments. Exceptions are Holocene channels, such as the Bastrop, Oyster Creek, Caney Creek, and Rio Grande distributary channels. Erosion of these paleochannels is occurring today and continues to be an important source of sand to the coast.

It is important to recognize that shoreface erosion is itself an important source of sand for coastal barriers. The thicker the barrier, the more sand is exhumed during transgression and recycled back into the coastal system. In the case of transgressive barriers, including Bolivar Peninsula, Follets Island, Matagorda Peninsula, and South Padre Island, these barriers have been reduced to a few meters of sand. Thus, less sand is being recycled back into these retreating barriers as transgression continues, which ultimately contributes to their demise. However, thicker barriers, such as Galveston Island, Matagorda Island, San Jose Island, and Mustang Island, are undergoing self-cannibalization that has slowed their rate of landward retreat.

Climate Change and Sediment Supply from Rivers. There were a number of global climate changes during the Holocene that may have impacted coastal evolution by influencing riverine sediment supply to the coast, wave climate and nearshore circulation, and the frequency and magnitude of severe storms. These include the Holocene Climate Optimum (ca. 9.0 to ca. 4.0 ka), when global temperatures were relatively warm, the Common Era (defined here as ca. 3.5 ka to ca. 1.2 ka), which was marked by cooling in the Northern Hemisphere, the relatively warm Medieval Climate Anomaly (ca. 1.2 ka to ca. 0.8 ka), and the Little Ice Age (ca. 1300 CE to ca. 1850 CE). The degree to which these events impacted coastal processes in the LaTex area remains uncertain, but it is logical that rivers with larger drainage basins, including the Mississippi River, were impacted by these global events while rivers with smaller drainage basins were more influenced by regional climate variability.

Within the LaTex area, the pace of climate change increased during the Holocene and was dominated by alternations between cool and wet and warm and dry conditions (Toomey et al., 1993; Nordt et al., 2002; Forman et al., 2009; Buzas-Stephens et al., 2014; Livsey et al., 2016). A recent Holocene paleoclimate review by Wong et al. (2015) shows considerable paleoclimate variability across Texas, with an overall trend from cool and wet conditions of the early Holocene transitioning to warmer and drier conditions in the middle Holocene, followed by more variable climate conditions across the region during the late Holocene (Fig. 54). However, the various studies used for their synthesis include both terrestrial and marine records and different paleoclimate proxies, so the record remains incomplete.

The Mississippi River has been, and continues to be, a significant sediment source for the LaTex Coast. Its modern suspended sediment discharge averages 200 10^6 t/yr at Vicksburg, Mississippi, but the amount of sediment reaching the coast has been significantly altered by humans. Estimates of the river's long-term discharge rate using the BQART calculation are ~428 10^6 t/yr (Syvitski and Milliman, 2007). This is significantly higher than the rate at the Rio Grande, which has the highest long-term sediment discharge of all Texas rivers (Fig. 3). Indeed, the Mississippi River has been the dominant source of silt and clay for the western Louisiana and Texas continental shelves. This includes the extensive and thick Texas Mud Blanket, which experienced a phase of significant growth during the late Holocene that is attributed to a change in climate and associated sediment discharge and dispersal (Weight et al., 2011). Late Holocene expansion of the Texas Mud Blanket is believed to have played a key role in the reversal from progradation to transgression of the central Texas Coast (Odezulu et al., 2020). Also, during the late Holocene, the Mississippi River was the dominant source of sand to the western Louisiana Chenier Plain. To date, the direct relationship between climate and sediment discharge from the Mississippi River remains problematic.

Because of their smaller drainage basin areas, rivers that drain directly into the LaTex Coast are considered to be sensitive to regional climate changes, such as those that occurred during the Holocene, in particular oscillations in precipitation as recorded by paleoclimate records (Fig. 54). As the rate of sea-level rise decreased during the middle and late Holocene, there continued to be changes in the size and extent of offshore deltas and bayhead deltas and in aggradation rates within the Brazos and Colorado River valleys. Figure 3 shows a strong correlation between drainage basin size and sediment discharge. The size, relief, and geology of the east Texas river drainage basins did not change significantly during the Holocene, so it is assumed that climate change played an important role in the observed changes in sediment discharge from rivers. We can test this assumption by relating the record of Holocene climate change (Fig. 54) to long-term sediment discharge estimates derived using the BQART method (Syvitski and Milliman, 2007) and sediment volume estimates from offshore deltas (Anderson et al., 2016) and bayhead deltas (Milliken et al., 2017; Fig. 3).

During the early Holocene, when the climate in the region was cool and wet (Fig. 54), the Brazos, Colorado and Rio Grande rivers maintained sizable offshore deltas. BQART calculations for the Colorado River Delta indicate significantly higher sediment discharge during the early Holocene relative to the middle late Holocene (Anderson et al., 2016). These changes occurred during the climate transition from cool and wet to warm and dry conditions (Fig. 54). From this time on, bayhead deltas across the region advanced and retreated in a diachronous manner, reflecting the variable climate and river sediment supply across the region (Fig. 55). This is consistent with results from Milliken et al. (2017), who examined climate-induced variability of Texas rivers using Geographic Information System (GIS) modeling and observational data and concluded that contemporaneous changes in the sediment yields across the region are unlikely given its highly variable climate setting. They further concluded that sediment yield is greatest in semiarid to subhumid fluvial systems, with diminishing sediment yield in arid or humid climatic regimes. In Galveston Bay, the Trinity Delta experienced ~8 km of progradation and aggraded up to 7 m between 2.6 ka and 1.6 ka (Fig. 13). Pollen records from the bay indicate that this was a time when the local climate was transitioning from more arid to more humid conditions (Ferguson et al., 2018b). During this same time interval, both the Sabine and Calcasieu bayhead deltas shifted northward (Figs. 15 and 17), and the Lavaca bayhead delta experienced no significant change (Fig. 34). These differences suggest that the Trinity River, with its larger drainage basin area that spans a fairly large precipitation gradient, is more responsive to climate change than the other rivers of the eastern LaTex region.

Results from Corpus Christi Bay and Baffin Bay provide strong evidence for the sensitivity of these bays to climate change (Simms et al., 2010; Buzas-Stephens et al., 2014; Livsey and Simms, 2016; Ferguson et al., 2018b; Rice et al., 2020). The Nueces bayhead delta of Corpus Christi Bay experienced flooding events at ca. 5.0 ka and ca. 3.2 ka and intervening periods of progradation at ca. 6.6 ka, ca. 3.8 ka, and ca. 2.2 ka that are linked to periods of warm/dry and cool/wet conditions, respectively (Rice et al., 2020). Similar climate-controlled environmental changes in Baffin Bay are indicated by down-core variations in foraminiferal assemblages with different salinity tolerances (Buzas-Stephens et al., 2014). Livsey and Simms (2016) later provided documentation that flooding events within Baffin Bay were caused by reductions in sediment supply to the bay since ca. 6 ka that resulted from drier conditions in the region and culminated in the current unique hypersaline conditions in the bay.

Climate change has also influenced river sediment discharge at decadal timescales. The Brazos River mouth was diverted in 1929, but before then it maintained a sizable wave-dominated delta (Fig. 31). This delta was eroded within a few decades following the 1929 diversion, and a new delta formed at the relocated river mouth. The initial growth of this modern delta was dominated by sand delivery from erosion of the pre-1929 delta, which was located ~10 km to the east. Early growth of the delta was followed by a phase of episodic growth that manifest as sand ridges separated by mud flats and expansion of the offshore portions of the delta (Rodriguez et al., 2000; Carlin and Dellapenna, 2014). A detailed study of the delta led to the observation that this episodic growth was regulated by large flood events, when significant volumes of fine sand were delivered to the delta within a period of months and appeared to be regulated by El Niño–Southern Oscillation cycles (Rodriguez et al., 2000; Fraticelli, 2006). Fraticelli (2006) showed that episodes of increased sand supply to the coast were associated with periods of increased river discharge during El Niño events that followed extended periods of drought, which occurred during La Niña events. Thus, sand supply from the Brazos River, and presumably the Colorado River since the two rivers have adjacent drainage basins within the same climate zones, is strongly regulated by climate variability over decadal timescales.

River Avulsion Influence on Sediment Supply. Both the Brazos and Colorado rivers have experienced multiple avulsions during the Holocene, with avulsions occurring at a frequency of ca. 2.4 ka. Based on observations from the 1929 Brazos River diversion, each avulsion results in a change in the location where sand is delivered to the coast, erosion of the old delta, and formation of a new delta. These changes occurred within decades after avulsion took place.

The distance of the avulsion node from the coast is controlled by the backwater length, which itself is governed by the regional relief and fluvial gradient, among other factors, and influences the distance between channels at the coast (Chatanantavet et al., 2012). Avulsion nodes for the Brazos River have been located within 80 km of the coast and resulted in river mouth locations located within a 20 km stretch of the coast. Avulsion nodes of the Colorado River are located farther inland and resulted in shifts of the river mouth spanning 70 km; thus, they have had a greater impact on the coast.

Tidal Inlet and Delta Sediment Sequestration. Tidal inlets and deltas can be significant sand sinks that remove sand from the longshore transport system as they evolve (e.g., FitzGerald et al., 2012; Nienhuis and Lorenzo-Trueba, 2019). Tidal inlets within incised river valleys of the western Louisiana and east Texas coasts have a history of deep incision and landward migration, resulting in large volumes of sand being sequestered within these valleys. Bolivar Peninsula provides a good example of sand sequestration through lateral accretion of the barrier (Fig. 20). As the rate of sea-level rise decreased during the middle and late Holocene, coastal barrier and chenier formation resulted in more narrow, deeper tidal inlets and diminished tidal influence in those inlets that occupy incised river valleys. As a result, tidal deltas decreased in size relative to early Holocene tidal deltas (Fig. 11). During the late Holocene, flood tidal deltas within Lake

Calcasieu, Sabine Lake, and Galveston Bay shrank in size as chenier plain and coastal barrier development decreased the size of inlets (Figs. 13, 15, and 17). During the same time, the ebb tidal deltas in these areas increased in size. Along the central Texas Coast, tidal deposits make up a significant portion of coastal barriers. For example, Mustang Island is composed largely of tidal deposits due to the relatively frequent shifts in inlet location within the Nueces valley (Fig. 45C).

Another good example of the role of tidal deltas in sequestering sand is the San Luis Tidal Delta. Increased rates of erosion of Galveston Island since ca. 1.8 ka have coincided with growth of this tidal delta and depletion of sand delivery to Follets Island (Wallace and Anderson, 2013). Wallace and Anderson (2013) estimate the long-term (ca. 2.1 ka to 0.2 ka) sand flux from Galveston Island to the tidal delta to be ~4700 m^3/yr. The flux has more than doubled (~10,000 m^3/yr) over the past 200 years in response to accelerated erosion of Galveston Island. Although the timing of Pass Cavallo formation and its growth rate is not as well constrained as for the San Luis Tidal Delta, it was the ultimate sink for sand eroded from Matagorda Peninsula. Platform borings through the western end of Matagorda Peninsula show that the inlet has migrated ~15 km to the west and deposited upwards of 15 m of sandy tidal deposits (Fig. 40).

Offshore Sand Sequestration and Storm Overwash. Offshore sand transport during storms has been argued to be a significant process (Hayes, 1967; Swift et al., 1985; Snedden et al., 1988), but more recent studies indicate that the volumes of sand actually removed from the beach and upper shoreface may be significantly smaller than previously thought. Wallace and Anderson (2013) studied sediment cores from the inner shelf offshore Galveston Island, which contain multiple thin sand and shell storm beds. Their sediment flux calculations show that the storm sand flux has decreased from the middle Holocene to the late Holocene, with this trend suggesting that hurricanes have been a minor yet constant contributor to offshore sand transport (Wallace and Anderson, 2013). They also noted that these storm beds are composed of very fine sand that is finer than the sand that occurs onshore and in the upper shoreface. This was also noted by Cole and Anderson (1982), Siringan and Anderson (1994), and Odezulu et al. (2020). Storm beds being composed of finer grained sand than what is found on the beach is consistent with observations that a significant portion of sand removed from the upper shoreface is delivered back to the longshore transport system within months to years (Morton and Sallenger, 2003; Goff et al., 2019). Thus, removal of sand from the coast during storms appears to be minor compared to the amount of sand that is sequestered in tidal deltas and washover deposits.

Current rates of shoreline erosion and overwash on eastern Bolivar Peninsula, Follets Island, Matagorda Peninsula, and South Padre Island are resulting in islands that are getting lower, narrower, and thinner through time. A number of researchers have made the case that these changes will exacerbate overwash through time and, coupled with the limited accommodation in their back-barrier bays and lagoons, is contributing to increased rates of washover and landward migration of these barriers (Pilkey and Davis, 1987; Stolper et al., 2005; Moore et al., 2010). Indeed, Follets Island, Matagorda Peninsula, and South Padre Island are currently losing a significant portion of their sand budgets through storm overwash. The best documentation for this comes from a study of Follets Island's overwash history by Odezulu et al. (2018). They calculate a historical (past 70 years) overwash flux of 15,200 m^3/yr compared to a long-term (past ~3000 years) overwash flux for Follets Island of 2300 m^3/yr (Wallace et al., 2010). This historical flux accounts for just over half of the total volume of sand produced by shoreline erosion during this time period. Similar flux calculations for the other transgressive barriers have not been attempted, but aerial photographs and sediment cores clearly indicate that washover is an important mode of sand sequestration in these areas. Overwash fluxes are expected to increase in the future in response to acceleration of sea-level rise and increased storm magnitudes.

Oceanographic Influences and the Texas Mud Blanket

During the late Holocene, the central Texas shoreface experienced a transgressive event that coincided with expansion of the Texas Mud Blanket. Odezulu et al. (2020) argue that the dramatic shift from progradation to transgression along the central Texas Coast at this time was caused by expansion of the mud blanket and burial of offshore sand sources. This argument is supported by results from other areas, such as the Amazon-Guianas Coast, where Amazon mud blankets trap fluvial sand that would otherwise provide a source of sand for nourishing and stabilizing the coast (Anthony et al., 2010). In addition, the high influx and viscosity of mud-rich waters attenuate wave energy and stimulate mud deposition, including moderate energy environments, as the mud blanket dewaters and becomes welded to the shoreline (Wells and Coleman, 1981; Wells, 1983). Additional cases of thick mud deposition in moderate-energy environments have been reported along the Suriname Coast, Louisiana Coast, Cayenne Coast, and southwest coast of India (Wells, 1983; Kineke et al., 1996; Anthony et al., 2010).

Dramatic growth of the Texas Mud Blanket during the late Holocene is attributed to an increase in westerly winds (Weight et al., 2011), which drive circulation on the shelf and regulate fine-grained sediment transport (Sionneau et al., 2008). This increase in sediment flux also occurred at the end of a prolonged warm and dry interval (Climate Optimum) that was followed by cool and wet conditions, which would have increased sediment erosion and discharge across the region (Weight et al., 2011).

Impacts from Severe Storms

Marsooli et al. (2019) argue that the effect of climate change on tropical cyclone climatology will likely be larger than the impact of sea-level rise for over 40% of Gulf of Mexico counties. Climate models indicate that the frequency of severe storms (categories 3–5) will increase in the future as Gulf of Mexico water temperatures continue to rise. Another prediction is that increasing aridity will result in greater evaporation and moisture content of the atmosphere, which could result in a 30–40% increase in precipitation resulting from storms in the Gulf of Mexico (Bruyère et al., 2017). Indeed, Hurricane Harvey of 2017, which was an unprecedented precipitation event, may be more typical of future storms. It struck the Texas Coast twice, the first time causing severe wind damage as the storm intensified rapidly over unusually warm Gulf of Mexico waters just before making landfall. The storm then consumed massive amounts of moisture as it circled around and traveled back out over the Gulf before making its second landfall, resulting in unprecedented rainfall and flooding in the greater Houston metropolitan area (Zhang et al., 2018).

Our understanding of the climatology of severe storms and their impacts is largely limited to historical observations. Longer time-scale observations are needed to test numerical models that attempt to predict severe storm response to climate change. Unfortunately, research focused on the Holocene record of severe storm landfalls along the LaTex Coast has been limited to just a few locations, and the Holocene climate record for the LaTex area lacks the necessary details for a high-resolution comparison of these records. Wallace and Anderson (2010) acquired a sedimentary record of the frequency of hurricane impacts spanning ca. 5.3 ka to ca. 0.9 ka using sediment cores from Laguna Madre. Their results showed an average storm impact probability of ~0.46%, which is similar to the values (~0.39%) obtained from comparable studies in western Florida and Lake Shelby, Alabama (Liu and Fearn, 1993, 2000). This suggests that there was little change in storm frequencies or trajectories across the Gulf Coast over these multi-millennial timescales, but decadal- to centennial-resolution records are needed for comparison with higher resolution climatic and oceanographic variability.

Recent inverse numerical techniques (summarized in Wallace and Woodruff, 2020) are aimed at analyzing the intensity and frequency of storms using paleostorm deposits. This is often accomplished by calibrating a sedimentary archive to modern events and measuring the grain size and distance sediment traveled for each paleostorm deposit (Wallace et al., 2014). For back-barrier settings, this requires detailed knowledge of seaward Holocene barrier island geomorphic changes. One limitation for selecting an appropriate archive is that long-term preservation is often lacking due to erosion or bioturbation, and therefore site selection for future high-resolution archives remains crucial (Wallace et al., 2014).

Geomorphic features associated with severe storms provide another geological record of storm impacts during the Holocene. The most notable features are large storm surge channels and washover deposits, which are widespread across the LaTex Coast. Our results indicate that these features were mostly formed during the early development of coastal barriers, when they were narrower and lower, which makes them more susceptible to storm breaching and overwash. In the case of transgressive barriers such as Follets Island, Matagorda Peninsula, and South Padre Island, they have been reduced in width and elevation to the point where breaching and storm washover is currently active. So, we must exercise caution when inferring storm magnitudes from geomorphic features without good knowledge of the age of these features and their associated coastal barriers.

The most startling geological record of severe storm impact is found on Bolivar Peninsula. There, an erosional event removed virtually the entire peninsula, leaving a significant stratigraphic hiatus that is marked by a prominent shell lag separating beach deposits older than ca. 1.2 ka from deposits younger than ca. 0.65 ka (Fig. 22). This event was also captured in a dramatic spike in dinoflagellate concentrations in sediment core TBHD5-1 from the upper part of Trinity Bay, which is indicative of a major increase in salinity that extended to the upper reaches of Galveston Bay (Ferguson et al., 2018a). The healing phase of the peninsula is marked by up to 1.7 km of beach ridge development on the western end of the peninsula, while the eastern half of the barrier remains mostly low (<1 m elevation) and vulnerable to storm inundation and breaching (Fig. 19).

Most of the sand eroded from the peninsula was transported back onto the barrier to form the modern beach ridge complex with a formation frequency of ~80 years. This frequency is similar to the rate of ridge formation for the Sabine Chenier Plain. Other coastal barriers, including Follets Island, Matagorda Peninsula, Mustang Island, and South Padre Island, all of which lack beach ridges, may have experienced major storm impacts during the times major transgressive surfaces were formed in these barriers, but the sand eroded from them ended up as washover deposits. Similar to the South Padre/Laguna Madre paleohurricane archive (Wallace and Anderson, 2010), these other backbarrier settings could yield a record of intense paleohurricane activity.

The reality is that the influence of global climate change on severe storms remains uncertain, so we are "learning as we go," especially when it comes to how upper atmospheric changes may influence storm intensification and trajectories. For example, scientists are just starting to recognize that de-icing of the Arctic Ocean influences the behavior of the jet stream (Francis and Vavrus, 2015), which in turn can influence storm tracks in the northern Gulf of Mexico.

Anthropogenic Impacts

Humans have occupied the LaTex Coast throughout the Holocene (Worrall, 2021), but their impact on the coast was minimal until the late 1800s, when extensive agricultural activity began. There is little question that humans have considerably influenced coastal environments since that time (Wallace et al., 2009). Damming of the Brazos, Colorado, and Rio Grande rivers and excessive water use from the Rio Grande have led to a reduction in riverine sand supply to the coast. The modern Rio Grande Delta has multiple distributary channels that are being actively reworked by waves to form barrier islands (Fig. 48). How much of the reduction in sediment supply is related to climate change and how much is due to anthropogenic influence is not known. The construction of numerous dams and weirs, coupled with large-scale municipal and agricultural water usage, has resulted in massive reduction of the river's outflow and sediment delivery to the Gulf of Mexico (Benke and Cushing, 2005).

In contrast, deforestation and agricultural activity within some river drainage basins have resulted in increased sediment supply, especially for smaller rivers. For example, the Trinity bayhead delta in Galveston Bay has experienced increased sediment supply and rapid growth during historical time that is attributed to changing practices of land use (McEwen, 1969; Phillips et al., 2004; Rodriguez et al., 2020). Other impacts include increased subsidence due to ground water extraction, with the most dramatic case occurring in the upper portions of Galveston Bay (Morton et al., 2006; Qu et al., 2015). The area experienced up to 2 m of subsidence within a few decades, which led to significant land loss during the 1970s (Morton et al., 2006). Construction of ship channels and associated jetties has altered the longshore sand transport system, resulting in increased erosion along some stretches of the coast and the growth of beaches that lie within the shadows of these features. For example, more than 1 km of beach has accreted to the east end of Galveston Island since jetties were constructed. This is sand that is no longer moving to the west and contributes to coastal erosion on west Galveston Island. The ship channels have altered the salinity structure of bays, and spoil banks constructed along channel margins disrupt normal wind-driven and tidal circulation, resulting in altered sediment and nutrient transport within these estuaries. The Texas City Dike, which occurs along the western shore of Galveston Bay at the juncture with West Bay (Fig. 12), has restricted tidal circulation and altered bay salinities and sediment transport within the bay system.

There are many other examples of human impacts, but the fact is that human-induced alteration of coastal systems through engineering pales in comparison, both in magnitude and geographic extent, to those changes that occurred during the Holocene, especially those that occurred during the early Holocene. Global climate change has already resulted in a return to early Holocene rates of sea-level rise, the western portion of the study area is experiencing more arid conditions, and warmer surface water temperatures in the Gulf of Mexico are fueling wetter and stronger hurricanes. All of these changes are having an impact on the LaTex Coast and are exacerbated by anthropogenic modifications to the coast. This is why we need to better understand the response of coastal environments to changes in climate and sea-level rise and highlight the need to incorporate global climate change predictions when planning for engineered coastal management projects.

Ongoing and Future Change

Wetland and Bay Impacts

Between 1932 and 2016, Louisiana lost more than 3000 km^2 of its coast due to reduced sediment delivery from the Mississippi River and its tributaries, coupled with increased rates of relative sea-level rise (Couvillion et al., 2017). Törnqvist et al. (2020) conclude that loss of Louisiana's remaining ~15,000 km^2 of marshland is inevitable, including not only the delta but the chenier plain as well. Higher subsidence rates for the delta tend to be compensated for by higher vertical accretion rates than those of the chenier plain. As a result, Jankowski et al. (2017) estimate that ~65% of the wetlands of the delta are able to keep pace with present-day relative sea-level rise compared to only 42% of wetlands in the chenier plain. Their work also showed that the chenier plain is characterized by considerable spatial variability in subsidence that, when coupled with modest rates of vertical accretion, leads to a high risk of inundation in some areas. One of these areas is the lower Sabine valley, which is currently experiencing a 23% (~72 km^2) conversion of subaerial environments to marshland and pond settings (Milliken et al., 2008b). This is a sharp reversal from the long-term trend; both the Calcasieu and Sabine Chenier plains have had a history of growth over the past ~3000 years; the Calcasieu Chenier Plain has prograded up to 6 km at an average rate of +2.0 m/yr.

Although the total area of coastal land loss for Texas is significantly less than for Louisiana, erosion is gnawing away at the coastal barriers, which serve as a critical line of defense between the low-lying coastal plain and the impacts of storms and rising sea level. Net shoreline retreat has occurred along 84% of the coast since 1930, resulting in an estimated land loss of 56 km^2 (Paine et al., 2012). Unlike most of coastal Louisiana, some of the most vulnerable coastal areas in Texas are densely populated and home to large industrial facilities, in addition to a myriad of delicate ecosystems.

Bays are especially vulnerable to accelerated sea-level rise and changes in sediment supply. Bayline erosion rates in Texas typically exceed 1 m/yr and locally are >4.0 m/yr (White et al., 2002). These rates vary locally, reflecting variations in bayline response to prevailing wind and wave directions (fetch), substrate conditions, relative sea-level rise, terrigenous sediment supply, and organic sediment accretion rates. Given the magnitude of historical bayline erosion, significant change is predicted. But, given the number of variables involved, predicting geographic variability in bayline change is challenging.

Bays within the study area fall into two general categories, shore-perpendicular bays/estuaries that occupy incised fluvial valleys where compaction of thick Holocene sediments results in subsidence rates of between 2.5 mm/yr and 3.3 mm/yr, and shore-parallel bays and lagoons that generally occupy interfluve areas where the Pleistocene surface typically occurs less than 2 m below the bay floor, indicating subsidence rates of <1 mm/yr (Fig. 7). Again, the number of data points is sparse, so these rates do not show the degree of geographic variability expected given the high relief on the Pleistocene surface and thickness and character of Holocene sediments (Fig. 9). It is also noteworthy that most of the modern relative sea-level records for coastal Texas come from tide gauge records located within incised valleys (Fig. 7). So, extrapolation of these rates to outlying areas can result in unreliable estimates of relative sea-level rise.

In larger bays, such as Galveston Bay, the fastest bayline erosion occurs where wind fetch is greatest (White et al., 2002).

Larger longitudinal bays, such as West Bay and East Bay, are also experiencing rapid bayline erosion, whereas bays that border transgressive barriers, such as East Bay, Christmas Bay, and east Matagorda Bay, have more stable shorelines due to nourishment by storm washover. Salt marshes are especially vulnerable to wind-generated wave erosion due to their low elevation and intertidal position. Inundation of coastal wetlands is known to depend on marsh aggradation rates. In areas where marsh aggradation rates have been determined, they are within a few millimeters of the current rate of sea-level rise, which is why an increase of only a few millimeters in the rate of rise is of considerable importance in predicting future wetlands loss. This is why there has been considerable research aimed at measuring these rates for different wetland environments (e.g., Morris et al., 2002; Wu et al., 2017; Horton et al., 2019). Unfortunately, relatively little work has been done in Texas, where variable climate and vegetation influence wetland accretion rates. One exception is a study of West Bay, which is part of the Galveston Bay Complex (Fig. 12) and an area of extreme rates of wetlands loss (Ravens et al., 2009). Ravens and his colleagues determined a historic sediment accretion rate of 2.0 mm/yr, which is below the current rate of relative sea-level rise for the area. Given the historical acceleration of sea-level rise, and predictions for sea-level rise this century, coastal wetlands will continue to experience significant inundation, especially in western Louisiana. In Texas, recent flood hazard assessment predicts that over 76 km^2 of the coast will be flooded by 2100 (Miller and Shirzaei, 2021).

Bayhead deltas are another coastal environment that is highly vulnerable to accelerated sea-level rise and changes in sediment supply, although they are not as vulnerable as patch and fringing saltmarshes because rivers provide the terrigenous sediment needed to maintain high accretion rates (Wu et al., 2020). This further implies that variations in fluvial sediment supply to bayhead deltas can significantly influence their stability. The response of bayhead deltas to sea-level rise during the early Holocene was one of rapid landward shifts (kilometers per century) during punctuated sea-level rise that at times exceeded 10 mm/yr. This is more than double the estimated vertical aggradation rate of ~4.7 mm/yr, based on the average sediment accumulation rate for bayhead deltas. At ca. 7.0 ka, the rate of eustatic rise decreased to 1.4 mm/yr, so the rate of relative rise within incised valleys was closer to the sediment aggradation rate. From this time on, bayhead deltas across the region have advanced and retreated in a more diachronous manner, which reflects the variable climate and river sediment supply across the region (Fig. 55). This variability appears to have increased in historical time, with some bayhead deltas having decreased in size while others have experienced growth that is likely due to increased sediment supply related to anthropogenic influences, in particular agricultural activity, within their drainage basins.

Estuaries of the low-gradient western LaTex Coast are especially vulnerable to sea-level rise this century. Sabine Lake will experience significant widening as adjacent low-lying wetlands are flooded (Milliken et al., 2008b), and the Calcasieu bayhead delta will shift to a location near Lake Charles, Louisiana (Milliken et al., 2008c). Although the river gradients are higher along the central Texas Coast, portions of the Nueces bayhead delta have retreated up to 250 m since the 1950s. This retreat is likely due to a combination of erosion of the abandoned portions of the delta and growth of tidal ponds through inundation.

In addition to the impacts of sea-level rise, bays of the LaTex region are expected to experience changes in sediment supply from rivers and streams. In particular, increased aridity will result in changing vegetation patterns within drainage basins, and river discharge is expected to decrease. Liénard et al. (2016) used a tolerance distribution model (TDM) to examine forest tolerances and vulnerabilities to anticipated climate change along the Gulf Coast, focusing on the implications of an anticipated increase in drought conditions across the region. Based on an ensemble of 17 climate change models to drive this TDM, they estimate that 18% of U.S. ecosystems are vulnerable to drought over the coming century, and western Louisiana and west Texas will experience a dramatic increase in tropical dry forests, similar to those of eastern Mexico. Given model results from Milliken et al. (2017), this translates into a reduction in sediment delivery from these rivers to bays that will likely result in a reduction in the size of bayhead deltas and associated wetlands.

Coastal Barriers

The Holocene evolution of coastal barriers and cheniers has varied considerably within the study area, despite the order of magnitude decrease in the rate of sea-level rise. This supports numerical models, which indicate that multiple factors influence barrier stability (e.g., Moore et al., 2010; Lorenzo-Trueba and Ashton, 2014; Murray and Moore, 2018). Important factors include wave climate, beach and offshore gradient, long-shore sand budget, wave climate, storm magnitude and frequency, the thickness and composition of shoreface deposits, barrier height and width, and backshore accommodation for washover deposition. This explains why coastal evolution has varied along the LaTex Coast. It also means that predicting future change is complicated.

Monitoring of LaTex shoreline migration extends back to the mid-nineteenth century with most locations showing a trend toward increasing, but highly variable, rates of landward shoreline movement (Paine et al., 2012, 2021). The Calcasieu and Sabine Chenier plains formed in the last ca. 3 ka and had a history of nearly continuous growth with average rates of +2.0 m/yr to +3.0 m/yr. Historical shoreline migration data for the area are sparse but indicate that the average rate of growth has decreased to about +0.8 m/yr during the past century. Portions of the chenier plain have experienced erosion, with rates ranging from −2.7 m/yr to −31.0 m/yr during multi-decadal periods between 1884 and 2005 (Martinez et al., 2009). In Texas, landward retreat of coastal barriers is occurring at rates that are unprecedented for the late Holocene (Fig. 56). This includes the more stable central Texas Coast, where historical shoreline migration rates are relatively small but where the coast was prograding prior to historical time. These are among the thickest barrier islands of the LaTex Coast, and their slow retreat is largely because they are being nourished by sand recycled from the barrier and shoreface during transgression. The most dramatic changes are occurring on Follets Island, Matagorda Peninsula, and South Padre Island. These transgressive barriers have beach and upper shoreface sands less than 1.5 m thick and rest on back-barrier bay, deltaic, or Pleistocene deposits. As these barriers migrate landward, only small amounts of sand are recycled landward from the eroding shoreface, further increasing rates of landward retreat. Furthermore, these transgressive barriers lack beach ridges and large eolian dunes, which results in lower elevations and more efficient overwash that further enhances landward retreat. Rates of barrier migration are also controlled by back-barrier accommodation (bay depth), which varies between ~1 m (Follets Island) to ~3 m (Matagorda Peninsula).

The historical changes to chenier plains and coastal barriers of the LaTex Coast are attributed to the depletion of sand supply and accelerated sea-level rise (Wallace and Anderson, 2013; Anderson et al., 2014; Odezulu et al., 2018). Direct human influence —such as alteration to water and sediment discharge from rivers and construction of hard structures—has become a key contributor to coastal change, but these impacts have been more localized than those caused by global climate change. Given these observations, the Texas Coast is trending toward more isolated barriers separated by wide stretches of strand plains, similar to the dry, sand-starved coast of northern Mexico (Davis, 2011).

CONCLUSIONS

Sea-level rise dominated coastal evolution in the LaTex region during the early Holocene, when the rate of rise in the northern Gulf of Mexico averaged 4.2 mm/yr and the overall rate was punctuated by brief periods of time when the rate exceeded 10 mm/yr. These events occurred at ca. 9.6 ka, ca. 8.8 ka, ca. 8.2 ka, and ca. 7.7 ka and resulted in significant landward shifts of deltas, coastal barriers, and bay environments on the scale of hundreds of meters to kilometers per century. The widespread nature of these events and their impact on different coastal environments indicate that they were not caused by autogenic mechanisms, which are expected to vary temporally and spatially. Rather, the episodic nature of sea-level rise and the magnitudes of these punctuated events is consistent with results from investigations that indicate episodes of rapid and large-scale retreat of portions of the Antarctic Ice Sheet during the early Holocene (Bentley et al., 2014; Prothro et al., 2020). They are also similar in magnitude to sea-level events that could result from large scale retreat of Antarctica's larger ice streams (e.g., Rignot, 2021).

By middle Holocene time (ca. 7.0 ka), the rate of sea-level rise decreased to an average rate of 1.4 mm/yr, and the rate and magnitude of coastal change decreased as climate change began to overshadow sea-level rise in controlling coastal evolution. The shoreline continued to migrate landward across the low-gradient western Louisiana and east Texas continental shelf while the central Texas shoreline began to stabilize and prograde, marking the early development of modern coastal barriers. These barriers were nourished mainly by sand eroded from the abandoned Brazos, Colorado, and Rio Grande deltas. Coastal bays continued to experience change, but the timing and rates of change varied along the coast, again indicating that climate change and associated changes in sediment supply to these bays dominated their evolution.

By ca. 4.0 ka, the average rate of sea-level rise had slowed to 0.5 mm/yr. Despite this slow rate of sea-level rise, coastal barriers and bays continued to experience significant change. These changes varied in magnitude and timing along the coast, indicating that factors other than sea-level rise dominated coastal evolution. Among these factors are antecedent topography, on which modern coastal environments were established, and variable sediment supply from rivers and offshore sources. Other factors included sequestration of sand in storm washover deposits, offshore storm beds, and tidal inlets and deltas. The late Holocene was also a time of continued climate change in Texas as the mid-Holocene Climate Optimum and associated widespread dry conditions that had persisted in central and south Texas transitioned to a more variable climate that oscillated between cool/wet and warm/dry conditions (Fig. 54). The timing and magnitude of these climate oscillations varied across the region, as did their impacts. For example, the Trinity bayhead delta prograded and aggraded as sediment supply to the delta outpaced sea-level rise during the same time that Baffin Bay succumbed to prolonged dry conditions and took on its current, unique hypersaline state. Starting at ca. 3.0 ka, the Calcasieu and Sabine Chenier plains experienced significant growth that gradually isolated Lake Calcasieu and Sabine Lake from the Gulf of Mexico. Likewise, Galveston Island and Bolivar Peninsula experienced growth and progradation. Meanwhile, the central Texas Coast experienced a long transgressive phase as the Texas Mud Blanket expanded and buried offshore sand sources and draped the lower shoreface in marine mud.

There is a sparse paleorecord of severe storm impacts across the Gulf Coast, but those data that do exist indicate little change in storm frequency during the late Holocene (Wallace and Anderson, 2010; Wallace et al., 2014). This being said, at ~650 years ago, there is a disturbing record of a large storm impact on Bolivar Peninsula that virtually destroyed the barrier and resulted in a salinity spike that extended into the upper reaches of Galveston Bay and persisted for more than a century.

In sharp contrast to the diachronous changes that shaped the LaTex Coast during the late Holocene, the current condition is one where shoreline retreat is occurring along virtually the entire LaTex Coast (Fig. 56). In fact, historical rates of coastal change are approaching those of the early Holocene. What is different between now and the early Holocene is that significantly less sand is now available to maintain the balance between sea-level rise and sediment supply. In particular, sand supply from the Brazos, Colorado, and Rio Grande rivers has decreased and erosion of offshore deltas and fluvial channels was significantly diminished as these offshore sand sources were bypassed by the advancing shoreline and buried beneath marine mud. Humans have also contributed to the reduction in sediment supply to the coast and its dispersal. Given this reduction in sand supply to the coast, the impact of the six-fold acceleration in the rate of sea-level rise, from an average late Holocene rate of 0.5 mm/yr to a current rate near 3.0 mm/yr, is strongly exacerbated. Indeed, this historical change in shoreline trajectory marks a return to early Holocene conditions.

While the early Holocene is a good analog for understanding current and predicted coastal response to sea-level rise, the middle and late Holocene provide the best analogs for climate-induced coastal change. This is especially true for Texas, where climate models predict more arid conditions (Nielsen-Gammon et al., 2020). Among the expected impacts are reductions in sediment supply to bays from rivers and stronger and wetter tropical storms. Direct human influence, such as alterations to water and sediment discharge from rivers and the construction of hard structures, has become a key contributor to coastal change, but the impacts have been more localized than those caused by global climate change.

Combating coastal change will require scientific understanding of those processes that have and continue to influence the coast. In this regard, Louisiana is incorporating science into its coastal planning strategy and acquiring data necessary for more reliable predictions of coastal change. Texas has been less aggressive in applying science-based knowledge to understand and predict the impacts of climate change and at developing strategies for dealing with these changes. In particular, there is a dire need for more coastal subsidence measurements, analysis of vertical accretion rates in different wetland settings, numerical modeling of barrier response to various forcing mechanisms along with field data needed to constrain these models, improved understanding of coastal sediment budgets and transport, and assessment of offshore sand resources for beach nourishment. Lastly, there is a need for educational outreach focused on the realities of global climate change and its impacts on coastal environments.

ACKNOWLEDGMENTS

This research spans nearly four decades, so it is not possible to acknowledge all of the students and colleagues who have contributed, but most have co-authored papers that are cited in this Memoir. Likewise, we have received funding from a number of sources over the years, most notably the National Science Foundation, the Gulf of Mexico Quaternary Seismic Stratigraphy Consortium, the David Worthington Fund, and the Shell Center for Sustainability. Reviews from Dr. Joan Florsheim and an anonymous reviewer were helpful and greatly appreciated.

The content solely represents the views of the authors and does not necessarily represent the official views of the U.S. Army Corps of Engineers.

REFERENCES CITED

Abbott, J.T., 2001, Houston Area Geoarchaeology: A Framework for Archeological Investigation, Interpretation, and Cultural Resource Management in the Houston Highway District: Texas Department of Transportation, Environmental Affairs Division, Archeological Studies Program Report 27, 235 p.

Abdulah, K.C., Anderson, J.B., Snow, J.B., and Holdford-Jack, L., 2004, The late Quaternary Brazos and Colorado Deltas, offshore Texas—Their evolution and the factors that controlled their deposition, Late Quaternary stratigraphic evolution of the northern Gulf of Mexico: A synthesis, in Anderson, J.B., and Fillon, R.H., eds., Late Quaternary Stratigraphic Evolution of the Northern Gulf of Mexico Basin: Society for Sedimentary Geology (SEPM) Special Publication 79, p. 237–270, https://doi.org/10.2110/pec.04.79.0237.

Al Mukaimi, M.E., Dellapenna, T.M., and Williams, J.R., 2018, Enhanced land subsidence in Galveston Bay, Texas: Interaction between sediment accumulation rates and relative sea level rise: Estuarine, Coastal and Shelf Science, v. 207, p. 183–193, https://doi.org/10.1016/j.ecss.2018.03.023.

Anderson, J.B., and Rodriguez, A.B., eds., 2008, Response of Gulf Coast Estuaries to Sea-Level Rise and Climate Change: Geological Society of America Special Paper 443, 146 p., https://doi.org/10.1130/SPE443.

Anderson, J.B., and Thomas, M.A., 1991, Marine ice sheet decoupling as a mechanism for rapid, episodic sea-level change: The record of such events and their influence on sedimentation: Sedimentary Geology, v. 70, p. 87–104, https://doi.org/10.1016/0037-0738(91)90136-2.

Anderson, J.B., Abdulah, K., Sarzalejo, S., Siringan, F., and Thomas, M.A., 1996, Late Quaternary sedimentation and high-resolution sequence stratigraphy of the East Texas shelf, in De Batist, M., and Jacobs, P., eds., Geology of Siliciclastic Seas: Geological Society, London, Special Publication 117, p. 95–124, https://doi.org/10.1144/GSL.SP.1996.117.01.06.

Anderson, J.B., Rodriguez, A.B., Abdulah, K., Banfield, L.A., Bart, P., Fillon, R., McKeown, H., and Wellner, J., 2004, Late Quaternary stratigraphic evolution of the northern Gulf of Mexico: A synthesis, in Anderson, J.B., and Fillon, R.H., eds., Late Quaternary Stratigraphic Evolution of the Northern Gulf of Mexico Basin: Society for Sedimentary Geology (SEPM) Special Publication 79, p. 1–24, https://doi.org/10.2110/pec.04.79.0001.

Anderson, J.B., Rodriguez, A.B., Milliken, K., and Taviani, M., 2008, The Holocene evolution of the Galveston Bay Complex, Texas: Evidence for rapid change in estuarine environments, in Anderson, J.B., and Rodriguez, A.B., eds., Response of Gulf Coast Estuaries to Sea-Level Rise and Climate Change: Geological Society of America Special Paper 443, p. 89–104, https://doi.org/10.1130/2008.2443(06).

Anderson, J.B., Wallace, D.J., Simms, A.R., Rodriguez, A.B., and Milliken, K.T., 2014, Variable response of coastal environments of the Northwestern Gulf of Mexico to sea-level rise and climate change: Implications for future change: Marine Geology, v. 352, p. 348–366, https://doi.org/10.1016/j.margeo.2013.12.008.

Anderson, J.B., Wallace, D.J., Simms, A.R., Rodriguez, A.B., Weight, R.W., and Taha, P.Z., 2016, The recycling of sediments between source and sink during a eustatic cycle: Earth-Science Reviews, v. 153, p. 111–138, https://doi.org/10.1016/j.earscirev.2015.10.014.

Anthony, E.J., Gardel, A., Gratiot, N., Proisy, C., Allison, M.A., Dolique, F., and Fromard, F., 2010, The Amazon-influenced muddy coast of South America: A review of mud-bank–shoreline interactions: Earth-Science Reviews, v. 103, v. 99–121, https://doi.org/10.1016/j.earscirev.2010.09.008.

Ashton, A.D., and Lorenzo-Trueba, J., 2018, Morphodynamics of barrier response to sea-level rise, in Moore, L.J., and Murray, A.B., eds., Barrier Island Dynamics and Response to Changing Climate: Cham, Switzerland, Springer, p. 277–304, https://doi.org/10.1007/978-3-319-68086-6_9.

Aslan, A., and Blum, M.D., 1999, Contrasting styles of Holocene avulsion, Texas Gulf Coastal Plain, USA, in Smith, N.D., and Rogers, J., eds., Fluvial Sedimentology VI: Oxford, UK, Blackwell Publishing Ltd., p. 193–209, https://doi.org/10.1002/9781444304213.ch15.

Aten, L.E., 1983, Indians of the Upper Texas Coast: New York, Academic Press, 370 p.

Bamber, J.L., Oppenheimer, M., Kopp, R.E., Aspinall, W.P., and Cooke, R.M., 2019, Ice sheet contributions to future sea-level rise from structured expert judgment: Proceedings of the National Academy of Sciences of the United States of America, v. 116, p. 11,195–11,200, https://doi.org/10.1073/pnas.1817205116.

Banfield, L., and Anderson, J.B., 2004, The Late Quaternary evolution of the Rio Grande Delta: Complex response to eustasy and climate change, in Anderson, J.B., and Fillon, R.H., eds., Late Quaternary Stratigraphic Evolution of the Northern Gulf of Mexico Basin: Society for Sedimentary

Geology (SEPM) Special Publication 79, p. 289–306, https://doi.org/10.2110/pec.04.79.0289.

Bard, E., Hamelin, B., and Fairbanks, R.G., 1990, U–Th ages obtained by mass spectrometry in corals from Barbados: Sea level during the past 130,000 years: Nature, v. 346, p. 456–458, https://doi.org/10.1038/346456a0.

Bard, E., Hamelin, B., Arnold, M., Montaggioni, L., Cabioch, G., Faure, G., and Rougerie, F., 1996, Deglacial sea-level record from Tahiti corals and the timing of global meltwater discharge: Nature, v. 382, p. 241–244, https://doi.org/10.1038/382241a0.

Bart, P.J., Krogmeier, B.J., Bart, M., and Tulaczyk, S., 2017, The paradox of a long grounding during West Antarctic Ice Sheet retreat in Ross Sea: Nature: Scientific Reports, v. 7, 1262, https://doi.org/10.1038/s41598-017-01329-8.

Behrens, E.W., 1963, Buried Pleistocene river valleys in Aransas and Baffin Bays, Texas: Publication of the Institute of Marine Science, v. 9, p. 7–18.

Behrens, E.W., 1966, Surface salinities for Baffin Bay and Laguna Madre, Texas: Contributions in Marine Science, v. 11, p. 168–173.

Belknap, D.F., and Kraft, J.C., 1981, Preservation potential of transgressive coastal lithosomes on the U.S. Atlantic Shelf: Marine Geology, v. 42, p. 429–442, https://doi.org/10.1016/0025-3227(81)90173-0.

Benke, A., and Cushing, C., 2005, Rivers of North America: Burlington, Massachusetts, Elsevier Academic Press, p. 375–424.

Bentley, M.J., Cofaigh, C.O., Anderson, J.B., Conway, H., Davies, B., Graham, A.G.C., Hillenbrand, C.D., Hodgson, D.A., Jamieson, S.S., Larter, R.D., Mackintosh, A., Smith, J.A., Verleyen, E., Ackert, R.P., Bart, P.J., Berg, S., Brunstein, Canals, M., Colhoun, E.A., Crosta, X., Dickens, W.A., Domack, E., Dowdeswell, J.A., Dunbar, R., Ehrmann, W., Evans, J., Favier, V., Fink, D., Fogwill, C.J., Glasser, N.F., Gohl, K., Golledge, N.R., Goodwin, I., Gore, D.B., Greenwood, S.L., Hall, B.L., Hall, K., Hedding, D.W., Hein, A.S., Hocking, E.P., Jakobsson, M., Johnson, J.S., Jomelli, V., Jones, R.S., Klages, J.P., Kristoffersen, Y., Kuhn, G., Leventer, A., Licht, K., Lilly, K., Lindow, J., Livingstone, S.J., Massé, G., McGlone, M.S., McKay, R.M., Melles, M., Miura, H., Mulvaney, R., Nel, W., Nitsche, F.O., O'Brien, P.E., Post, A.L., Roberts, S.J., Saunders, K.M., Selkirk, P.M., Simms, A.R., Spiegel, C., Stolldorf, T.D., Sugden, D.E., van der Putten, N., van Ommen, T., Verfaillie, D., Vyverman, W., Wagner, B., White, D.A., Witus, A.E. and Zwartz, D., 2014, A community-based geological reconstruction of Antarctic Ice Sheet deglaciation since the Last Glacial Maximum: Quaternary Science Reviews, v. 100, p. 1–9, https://doi.org/10.1016/j.quascirev.2014.06.025.

Bernard, H.A., Major, C.F., Jr., and Parrott, B.S., 1959, The Galveston barrier island and environments: A model for predicting reservoir occurrence and trend: Transactions of the Gulf Coast Association of Geological Societies, v. 9, p. 221–224.

Bernard, H.A., Major, C.F., Jr., Parrott, B.S., and Leblanc, R.J., Sr., 1970, Recent Sediments of Southeast Texas: A Field Guide to the Brazos Alluvial and Deltaic Plains and the Galveston Barrier Island Complex: The University of Texas at Austin, Bureau of Economic Geology Guidebook 11, 132 p.

Berryhill, H.L., 1987, The continental shelf off south Texas, in Berryhill, H.L., ed., Late Quaternary Facies and Structure, Northern Gulf of Mexico: Interpretations from Seismic Data: American Association of Petroleum Geologists, Studies in Geology, v. 23, p. 11–79.

Blum, M.D., 1993, Genesis and architecture of incised valley fill sequences: A late Quaternary alluvial plain deposition, Texas coastal plain: Gulf Coast Association of Geological Societies Transactions, v. 44, p. 85–92.

Brain, M.J., Kemp, A.C., Horton, B.P., Culver, S.J., Parnell, A.C., and Cahill, N., 2015, Quantifying the contribution of sediment compaction to late Holocene salt-marsh sea-level reconstructions, North Carolina, USA: Quaternary Research, v. 83, p. 41–51, https://doi.org/10.1016/j.yqres.2014.08.003.

Brenner, O.T., Moore, L.J., and Murray, A.B., 2015, The complex influences of back-barrier deposition, substrate slope and underlying stratigraphy in barrier island response to sea- level rise: Insights from the Virginia Barrier Islands, Mid-Atlantic Bight, U.S.A.: Geomorphology, v. 246, p. 334–350, https://doi.org/10.1016/j.geomorph.2015.06.014.

Bruyère, C.L., Rasmussen, R., Gutmann, E., Done, J., Tye, M., Jaye, A., Prein, A., Mooney, P., Ge, M., Fredrick, S., Friis-Hansen, P., Garrem, L., Veldore, V., and Niesel, J., 2017, Impact of Climate Change on Gulf of Mexico Hurricanes: National Center for Atmospheric Research Technical Note NCAR/TN-535+STR, 165 p., https://doi.org/10.5065/D6RN36J3.

Buzas-Stephens, P., Livsey, D.N., Simms, A.R., and Buzas, M.A., 2014, Estuarine foraminifera record Holocene stratigraphic changes and Holocene climate changes in ENSO and the North American monsoon: Baffin Bay, Texas: Palaeogeography, Palaeoclimatology, Palaeoecology, v. 404, p. 44–56, https://doi.org/10.1016/j.palaeo.2014.03.031.

Byrne, J.R., 1975, Holocene depositional history of Lavaca Bay, Central Texas Gulf Coast [Ph.D. thesis]: Austin, Texas, University of Texas at Austin, 149 p.

Byrne, J.V., LeRoy, D.O., and Riley, C.M., 1959, The chenier plain and its stratigraphy, southwestern Louisiana: Transactions of the Gulf Coast Association of Geological Societies, v. 9, p. 237–260.

Carlin, J., and Dellapenna, T., 2014, The evolution of a subaqueous delta in the Anthropocene: A stratigraphic investigation of the Brazos River Delta, TX USA: Continental Shelf Research, v. 111, p. 139–149, https://doi.org/10.1016/j.csr.2015.08.008.

Carlin, J., Dellapenna, T., Figlus, J., and Harter, C., 2015, Investigating morphological and stratigraphic changes to the submarine shoreface of a transgressive Barrier Island: Follets Island, Northern Gulf of Mexico, in Wang, P., Rosati, J.D., and Cheng, J., eds., Coastal Sediments: The Proceedings of the Coastal Sediments 2015: Singapore, World Scientific, https://doi.org/10.1142/9789814689977_0004.

Carlin, J.A., Schreiner, K.M., Dellapenna, T.M., McGuffin, A., and Smith, R.W., 2021, Evidence of recent flood deposits within a distal shelf depocenter and implications for terrestrial carbon preservation in non-deltaic shelf settings: Marine Geology, v. 431, https://doi.org/10.1016/j.margeo.2020.106376.

Caudle, T.L., Paine, J.G., Andrews, J.R., and Saylam, K., 2019, Beach, dune, and nearshore analysis of southern Texas Gulf Coast using Chiroptera LIDAR and Imaging System: Journal of Coastal Research, v. 35, p. 251–268, https://doi.org/10.2112/JCOASTRES-D-18-00069.1.

Cazenave, A., Dominh, K., Guinehut, S., Berthier, E., Llovel, W., Ramillien, G., Ablain, M., and Larnicol, G., 2009, Sea level budget over 2003–2008: A reevaluation from GRACE space gravimetry, satellite altimetry and Argo: Global and Planetary Change, v. 65, p. 83–88, https://doi.org/10.1016/j.gloplacha.2008.10.004.

Chan, A.W., and Zoback, M.D., 2007, The role of hydrocarbon production on land subsidence and fault reactivation in the Louisiana coastal zone: Journal of Coastal Research, v. 233, p. 771–786, https://doi.org/10.2112/05-0553.

Chappell, J., Omura, A., Esat, T., McCulloch, M., Pandilfi, J., Ota, Y., and Pillans, B., 1996, Reconciliation of late Quaternary sea levels derived from coral terraces at Huon Peninsula with deep sea oxygen isotope records: Earth Planetary Science Letters, v. 141, p. 227–236, https://doi.org/10.1016/0012-821X(96)00062-3.

Chatanantavet, P., Lamb, M.P., and Nittrouer, J.A., 2012, Backwater controls of avulsion location on deltas: Geophysical Research Letters, v. 39, L01402, https://doi.org/10.1029/2011GL050197.

Chen, X., Zhang, X., Church, J.A., Watson, C.S., King, M.A., Monselesan, D., Legresy, B., and Harig, C., 2017, The increasing rate of global mean sea-level rise during 1993–2014: Nature Climate Change, v. 7, p. 492–495, https://doi.org/10.1038/nclimate3325.

Ciarletta, D.J., Lorenzo-Trueba, J., and Ashton, A.D., 2019, Interaction of sea-level pulses with periodically retreating barrier islands: Frontiers in Earth Science, https://doi.org/10.3389/feart.2019.00279.

Clark, P.U., Dyke, A.S., Shakun, J.D., Carlson, A.E., Clark, J., Wohlfarth, B., Mitrovica, J.X., Hostetler, S.W., and McCabe, A.M., 2009, The last glacial maximum: Science, v. 325, p. 710–714, https://doi.org/10.1126/science.1172873.

Cole, M.L., and Anderson, J.B., 1982, Detailed grain-size and heavy mineralogy of sands of the northeastern Texas Gulf Coast: Implications with regard to coastal barrier development: Transactions of the Gulf Coast Geological Society, v. 32, p. 555–563.

Coleman, J.M., and Roberts, H.H., 1988a, Sedimentary development of the Louisiana continental shelf related to sea level cycles: Part I—Sedimentary sequences: Geo-Marine Letters, v. 8, p. 63–108, https://doi.org/10.1007/BF02330967.

Coleman, J.M., and Roberts, H.H., 1988b, Sedimentary development of the Louisiana continental shelf related to sea level cycles: Part II—Seismic response: Geo-Marine Letters, v. 8, p. 109–119, https://doi.org/10.1007/BF02330968.

Couvillion, B.R., Beck, H., Schoolmaster, D., and Fischer, M., 2017, Land Area Change in Coastal Louisiana 1932 to 2016: U.S. Geological Survey Scientific Investigations Map 3381, 16 p., https://pubs.er.usgs.gov/publication/sim3381.

Cowell, P.J., and Kinsela, M.A., 2018, Shoreface controls on barrier evolution and shoreline change, in Moore, L.J., and Murray, A.B., eds., Barrier

Island Dynamics and Response to Changing Climate: Cham, Switzerland, Springer, p. 243–275, https://doi.org/10.1007/978-3-319-68086-6_8

Curray, J.R., 1960, Sediments and history of Holocene transgression, continental shelf, northwestern Gulf of Mexico, in Shepard, F.P., Phleger, F.B., and van Andel, T.H., eds., Recent Sediments, Northwestern Gulf of Mexico: Tulsa, Oklahoma, American Association of Petroleum Geologists, p. 221–266.

Dalrymple, D.W., 1964, Recent sedimentary facies of Baffin Bay, Texas [Ph.D. thesis]: Houston, Texas, Rice University, 192 p.

Dangendorf, S., Hay, C., Calafat, F.M., Marcos, M., Piecuch, C.G., Berk, K., and Jensen, J., 2019, Persistent acceleration in global sea-level rise since the 1960s: Nature Climate Change, v. 9, p. 705–710, https://doi.org/10.1038/s41558-019-0531-8.

Davis, R.A., 2011, Sea-Level Change and the Gulf of Mexico: College Station, Texas, Texas A&M University Press, 172 p.

Deschamps, P., Nicolas, D., Bard, E., Hamelin, B., Camoin, G., Thomas, A., Henderson, G.M., Okuno, J., and Yokoyama, Y., 2012, Ice-sheet collapse and sea-level rise at the Bølling warming 14,600 years ago: Nature, v. 483, p. 559–564, https://doi.org/10.1038/nature10902.

Donoghue, J.F., 2011, Sea level history of the northern Gulf of Mexico coast and sea level rise scenarios for the near future: Climatic Change, v. 107, p. 17, https://doi.org/10.1007/s10584-011-0077-x.

Driese, S.G., Nordt, L.C., Lynn, W.C., Stiles, C.A., Mora, C.I., and Wilding, L.P., 2005, Distinguishing climate in the soil record using chemical trends in a vertisol climosequence from the Texas coast prairie, and application to interpreting Paleozoic paleosols in the Appalachian Basin, U.S.A.: Journal of Sedimentary Research, v. 75, p. 339–349, https://doi.org/10.2110/jsr.2005.027.

Eckles, B., Fassell, M., and Anderson, J.B., 2004, Late Quaternary Evolution of the wave-storm-dominated Central Texas Shelf, in Anderson, J.B., and Fillon, R.H., eds., Late Quaternary Stratigraphic Evolution of the Northern Gulf of Mexico Basin: Society for Sedimentary Geology (SEPM) Special Publication 79, p. 271–288, https://doi.org/10.2110/pec.04.79.0271.

Englehardt, S.E., Horton, B.P., and Kemp, A.C., 2011, Holocene sea level changes along the United States' Atlantic Coast: Oceanography, v. 24, p. 70–79.

Ewing, T.E., and Gonzalez, J.L., 2016, The late Quaternary Rio Grande Delta—A distinctive, underappreciated geologic system: Transactions of the Gulf Coast Association of Geological Societies, v. 66, p. 169–180.

Fairbanks, R.G., 1989, A 17,000-year glacio-eustatic sea level record: Influence of glacial melting rates on the Younger Dryas event and deep-ocean circulation: Nature, v. 342, p. 637–642, https://doi.org/10.1038/342637a0.

Feagin, R.A., Yeager, K.M., Brunner, C.A., and Paine, J.G., 2013, Active fault motion in a coastal wetland: Matagorda, Texas: Geomorphology, v. 199, p. 150–159, https://doi.org/10.1016/j.geomorph.2012.08.013.

Ferguson, S., Warny, S., Anderson, J.B., Simms, A.R., and White, C., 2018a, Breaching of Mustang Island in response to the 8.2 ka sea-level event and impact on Corpus Christi Bay, Gulf of Mexico: Implications for future coastal change: The Holocene, v. 28, p. 166–172, https://doi.org/10.1177/0959683617715697.

Ferguson, S., Warny, S., Anderson, J.B., Simms, A.R., and Escaquel, G., 2018b, Holocene vegetation and climate evolution of Corpus Christi and Trinity bays: Implications on source-to-sink deposition on the Texas coast: Geobios, v. 51, p. 123–135, https://doi.org/10.1016/j.geobios.2018.02.007.

Fisk, H.N., 1948, Geological Investigation of the Lower Mermentau River Basin and Adjacent Areas in Coastal Louisiana: Vicksburg, Mississippi, Mississippi River Commission, U.S. Army Corps of Engineers, 40 p. and 26 plates.

FitzGerald, D.M., Fenster, M.S., Argow, B.A., and Buynevich, I.V., 2008, Coastal impacts due to sea-level rise: Annual Review of Earth and Planetary Sciences, v. 36, p. 601–647, https://doi.org/10.1146/annurev.earth.35.031306.140139.

FitzGerald, D.M., Buynevich, I.V., and Hein, C., 2012, Morphodynamics and facies architecture of tidal inlets and tidal deltas, in Davis, R.A., Jr., and Dalrymple, R.W., eds., Principles of Tidal Sedimentology: Cham, Switzerland, Springer, p. 301–333, http://doi.org/10.1007/978-94-007-0123-6_12.

Forbes, M.J., 1988, Hydrologic Investigation of Lower Calcasieu River Louisiana: U.S. Geological Survey Water Resources Investigations Report 87-4173, 61 p.

Forman, S.L., Nordt, L., Gomez, J., and Pierson, J., 2009, Late Holocene dune migration on the south Texas sand sheet: Geomorphology, v. 108, p. 159–170, https://doi.org/10.1016/j.geomorph.2009.01.001.

Francis, J.A., and Vavrus, S.J., 2015, Evidence for a wavier jet stream in response to rapid Arctic warming: Environmental Research Letters, v. 10, https://doi.org/10.1088/1748-9326/10/1/014005.

Fraticelli, C.M., 2006, Climate forcing in a wave-dominated delta: The effects of drought-flood cycles on delta progradation: Journal of Sedimentary Research, v. 76, p. 1067–1076, https://doi.org/10.2110/jsr.2006.097.

Galloway, W.E., Whiteaker, T.L., and Ganey-Curry, P., 2011, History of Cenozoic North American drainage basin evolution, sediment yield, and accumulation in the Gulf of Mexico Basin: Geosphere, v. 7, p. 938–973, https://doi.org/10.1130/GES00647.1.

Garrison, J.R., Jr., Williams, J., Potter Miller, S., Weber, E.T., McMechan, G., and Zeng, X., 2010, Ground-penetrating radar study of North Padre Island: Implications for barrier island internal architecture, model for growth of progradational microtidal barrier islands, and Gulf of Mexico sea-level cyclicity: Journal of Sedimentary Research, v. 80, p. 303–319, https://doi.org/10.2110/jsr.2010.034.

Goff, J.A., Lugrin, L., Gulick, S.P., Thirumalai, K., and Okumura, Y., 2016, Oyster reef die-offs in stratigraphic record of Corpus Christi Bay, Texas, possibly caused by drought-driven extreme salinity changes: The Holocene, v. 26, p. 511–519, https://doi.org/10.1177/0959683615612587.

Goff, J.A., Swartz, J.M., Gulick, S.P., Dawson, C.N., and de Alegria-Arzaburu, A.R., 2019, An outflow event on the left side of Hurricane Harvey: Erosion of barrier sand and seaward transport through Aransas Pass, Texas: Geomorphology, v. 334, p. 44–57, https://doi.org/10.1016/j.geomorph.2019.02.038.

Gould, H.R., and McFarlan, E., 1959, Geologic history of Chenier Plain, southwestern Louisiana: Transactions of the Gulf Coast Association of Geological Societies, v. 9, p. 261–270.

Greene, D.L., Jr., Rodriguez, A.B., and Anderson, J.B., 2007, Seaward-branching coastal plain and piedmont incised valley systems through multiple sea level cycles: Late Quaternary examples from Mobile Bay and Mississippi Sound, U.S.A.: Journal of Sedimentary Research, v. 77, p. 139–158, https://doi.org/10.2110/jsr.2007.016.

Hayes, M.O., 1967, Hurricanes as geological agents, south Texas coast: American Association of Petroleum Geologists Bulletin, v. 51, p. 937–942.

Hijma, M.P., Shen, Z., Torbjörn, T., and Mauz, B.E., 2017, Late Holocene evolution of a coupled, mud-dominated delta plain–chenier plain system, coastal Louisiana, USA: Earth Surface Dynamics, v. 5, p. 689–710, https://doi.org/10.5194/esurf-5-689-2017.

Hollis, R.J., Wallace, D.J., Miner, M.D., Gal, N.G., Dike, C., and Flocks, J.G., 2019, Late Quaternary evolution and stratigraphic framework influence on coastal systems along the north-central Gulf of Mexico, USA: Quaternary Science Reviews, v. 223, https://doi.org/10.1016/j.quascirev.2019.105910.

Horton, B.P., Shennan, I., Bradley, S.L., Cahill, N., Kirwan, M., Kopp, R.E., and Shaw, T.A., 2019, Predicting marsh vulnerability to sea-level rise using Holocene relative sea-level data: Nature Communications, v. 9, 2687, https://doi.org/10.1038/s41467-018-05080-0.

Howe, H.V.W., Russell, R.J., and McGuirt, J.H., 1935, Geology of Cameron and Vermilion Parishes; Physiography of Coastal Southwest Louisiana: Louisiana Geological Survey Bulletin 6, 68 p.

Humphrey, J.D., and Ferring, C.R., 1994, Stable isotopic evidence for latest Pleistocene and Holocene climatic change in north-central Texas: Quaternary Research, v. 41, p. 200–213, https://doi.org/10.1006/qres.1994.1022.

Israel, A.M., Ethridge, F.G., and Estes, E.L., 1987, A sedimentologic description of a micro-tidal, flood-tidal delta, San Luis Pass, Texas: Journal of Sedimentary Petrology, v. 47, p. 288–300.

Jankowski, K.L., Törnqvist, T.E., and Fernandes, A.M., 2017, Vulnerability of Louisiana's coastal wetlands to present-day rates of relative sea-level rise: Nature Communications, v. 8, https://doi.org/10.1038/ncomms14792.

Kemp, A.C., Horton, B.P., Donnelly, J.P., Mann, M.E., Vermeer, M., and Rahmstorf, S., 2011, Climate related sea-level variations over the past two millennia: Proceedings of the National Academy of Sciences of the United States of America, v. 108, p. 11,017–11,022, https://doi.org/10.1073/pnas.1015619108.

Kendall, R.A., Mitrovica, J.X., Milne, G.A., Törnqvist, T.E., and Li, Y., 2008, The sea-level fingerprint of the 8.2 ka climate event: Geology, v. 36, p. 423–426, https://doi.org/10.1130/G24550A.1.

Khanna, P., Droxler, A.W., Nittrouer, J.A., Tunnell, J.W., Jr., and Shirley, T.C., 2017, Coralgal reef morphology records punctuated sea-level rise during the last deglaciation: Nature Communications, v. 8, 1046, https://doi.org/10.1038/s41467-017-00966-x.

Kibler, K.W., 1999, Radiocarbon dating of *Rangia cuneata*: Correction factors and calibrations for the Galveston Bay area: Bulletin of the Texas Archeological Society, v. 70, p. 457–466.

Kineke, G.C., Sternberg, R.W., Trowbridge, J.H., and Geyer, W.R., 1996, Fluid-mud processes on the Amazon continental shelf: Continental Shelf Research, v. 16, p. 667–696, https://doi.org/10.1016/0278-4343(95)00050-X.

Knutson, T.R., McBride, J.L., Chan, J., Emanuel, K., Holland, G., Landsea, C., Held, I., Kossin, J.P., Srivastava, A., and Sugi, M., 2010, Tropical cyclones and climate change: Nature Geoscience, v. 3, p. 157–163, https://doi.org/10.1038/ngeo779.

Kolker, A.S., Allison, M.A., and Hameed, S., 2011, An evaluation of subsidence rates and sea-level variability in the northern Gulf of Mexico: Geophysical Research Letters, v. 38, L21404, https://doi.org/10.1029/2011GL049458.

Kuchar, J., Milne, G., Wolstencroft, M., Love, R., Tarasov, L., and Hijma, M., 2018, The influence of sediment isostatic adjustment on sea level change and land motion along the U.S. Gulf Coast: Journal of Geophysical Research: Solid Earth, v. 123, p. 780–796, https://doi.org/10.1002/2017JB014695.

Labeyrie, L.D., Duplessy, J.C., and Blanc, P.L., 1987, Variations in mode of formation and temperature of oceanic deep waters over the past 125,000 years: Nature, v. 327, p. 477–482, https://doi.org/10.1038/327477a0.

Lentz, E.E., Hapke, C.J., Stockdon, H.F., and Hehre, R.E., 2013, Improving understanding of near-term barrier island evolution through multi-decadal assessment of morphologic change: Marine Geology, v. 337, p. 125–139, https://doi.org/10.1016/j.margeo.2013.02.004.

Li, X.-X., Törnqvist, T.M., Nevitt, J.M., and Kohl, B., 2012, Synchronizing a sea-level jump, final Lake Agassiz drainage, and abrupt cooling 8200 years ago: Earth and Planetary Science Letters, v. 315–316, p. 41–50, https://doi.org/10.1016/j.epsl.2011.05.034.

Liénard, J., Harrison, J., and Strigul, N., 2016, US forest response to projected climate-related stress: A *tolerance* perspective: Global Change Biology, v. 22, p. 2875–2886, https://doi.org/10.1111/gcb.13291.

Liu, K.B., and Fearn, M.L., 1993, Lake-sediment record of late Holocene hurricane activities from coastal Alabama: Geology, v. 21, p. 793–796, https://doi.org/10.1130/0091-7613(1993)021<0793:LSROLH>2.3.CO;2.

Liu, K.B., and Fearn, M.L., 2000, Reconstruction of prehistoric landfall frequencies of catastrophic hurricanes in northwestern Florida from lake sediment records: Quaternary Research, v. 54, p. 238–245, https://doi.org/10.1006/qres.2000.2166.

Livsey, D., and Simms, A.R., 2013, Holocene sea-level change derived from microbial mats: Geology, v. 41, p. 971–974, https://doi.org/10.1130/G34387.1.

Livsey, D.N., and Simms, A.R., 2016, Episodic flooding of estuarine environments in response to drying climate over the last 6,000 years in Baffin Bay, Texas: Marine Geology, v. 381, p. 142–162, https://doi.org/10.1016/j.margeo.2016.09.003.

Livsey, D.N., Simms, A.R., Hangsterfer, A., Nisbet, R.A., and DeWitt, R., 2016, Drought modulated by North Atlantic sea surface temperatures for the last 3,000 years along the northwestern Gulf of Mexico: Quaternary Science Reviews, v. 135, p. 54–64, https://doi.org/10.1016/j.quascirev.2016.01.010.

Lohse, E.A., 1955, Dynamic geology of the modern coastal region, northwestern Gulf of Mexico, *in* Hough, J.L., and Menard, H.W., eds., Finding Ancient Shorelines: Society of Economic Mineralogists and Paleontologists (SEPM) Special Publication 3, p. 99–105, https://doi.org/10.2110/pec.55.01.0099.

Lorenzo-Trueba, J., and Ashton, A.D., 2014, Rollover, drowning, and discontinuous retreat: Distinct modes of barrier response to sea-level rise arising from a simple morphodynamic model: Journal of Geophysical Research: Earth Surface, v. 119, p. 779–801, https://doi.org/10.1002/2013JF002941.

Maddox, J., Anderson, J.B., Milliken, K., and Rodriguez, A.B., 2008, The Holocene evolution of the Matagorda and Lavaca estuary complex, Texas, USA, *in* Anderson, J.B., and Rodriguez, A.B., eds., Response of Gulf Coast Estuaries to Sea-Level Rise and Climate Change: Geological Society of America Special Paper 443, p. 105–119, https://doi.org/10.1130/2008.2443(07).

Marsooli, R., Lin, N., Emanuel, K., and Feng, K., 2019, Climate change exacerbates hurricane flood hazards along US Atlantic and Gulf Coasts in spatially varying patterns: Nature Communications, v. 10, 3785, https://doi.org/10.1038/s41467-019-11755-z.

Martinez, L., O'Brien, S., Bethel, M., Penland, S., and Kulp, M., 2009, Louisiana Barrier Island Comprehensive Monitoring Program (BICM) Volume 2: Shoreline Changes and Barrier Island Land Loss 1800's–2005 Report: Ponchartrain Institute Reports and Studies, Paper 1, 32 p.

McBride, R.A., Anderson, L.C., Tudoran, A., and Roberts, H.H., 1999, Holocene stratigraphic architecture of a sand-rich shelf and the origin of linear shoals: Northeastern Gulf of Mexico, *in* Bergman, K.M., and Snedden, J.W., eds., Isolated Shallow Marine Sand Bodies: Sequence Stratigraphic Analysis and Sedimentologic Interpretation: Society for Sedimentary Geology (SEPM) Special Publication 64, p. 95–126, https://doi.org/10.2110/pec.99.64.0095.

McBride, R.A., Taylor, M.J., and Byrnes, M.R., 2007, Coastal morphodynamics and Chenier-Plain evolution in southwestern Louisiana, USA: A geomorphic model: Geomorphology, v. 88, p. 367–422, https://doi.org/10.1016/j.geomorph.2006.11.013.

McCulloh, R.P., and Heinrich, P.V., 2012, Surface faults of the south Louisiana growth-fault province, *in* Cox, R.T., Tuttle, M.P., Boyd, O.S., and Locat, J., eds., Recent Advances in North American Paleoseismology and Neotectonics East of the Rockies: Geological Society of America Special Paper 493, p. 37–49, https://doi.org/10.1130/2012.2493(03).

McEwen, M.C., 1969, Sedimentary facies of the modern Trinity Delta, *in* Lankford, R.R., and Rogers, J.J.W., eds., Holocene Geology of the Galveston Bay Area: Houston, Texas, Houston Geological Society, p. 53–77.

McGowen, J.H., and Brewton, J.L., 1975, Historical Changes and Related Coastal Processes, Gulf and Mainland Shorelines, Matagorda Bay Area, Texas: Austin, Texas, The University of Texas at Austin, Bureau of Economic Geology Special Report SR0003, 72 p.

Meckel, T.A., ten Brink, U.S., and Williams, S.J., 2006, Current subsidence rates due to compaction of Holocene sediments in southern Louisiana: Geophysical Research Letters, v. 33, L11403, https://doi.org/10.1029/2006GL026300.

Mellett, C.L., and Plater, A.J., 2018, Drowned barriers as archives of coastal-response to sea-level rise, *in* Moore, L.J., and Murray, A.B., eds., Barrier Island Dynamics and Response to Changing Climate: Cham, Switzerland, Springer, p. 57–89, https://doi.org/10.1007/978-3-319-68086-6_2.

Miller, M.M., and Shirzaei, M., 2021, Assessment of future flood hazards for southeastern Texas: Synthesizing subsidence, sea-level rise, and storm surge scenarios: Geophysical Research Letters, v. 48, e2021GL092544, https://doi.org/10.1029/2021GL092544.

Milliken, K.T., Anderson, J.B., and Rodriguez, A.B., 2008a, A new composite Holocene sea-level curve for the northern Gulf of Mexico, *in* Anderson, J.B., and Rodriguez, A.B., eds., Response of Gulf Coast Estuaries to Sea-Level Rise and Climate Change: Geological Society of America Special Paper 443, p. 1–11, https://doi.org/10.1130/2008.2443(01).

Milliken, K.T., Anderson, J.B., and Rodriguez, A.B., 2008b, Tracking the Holocene evolution of Sabine Lake through the interplay of eustasy, antecedent topography, and sediment supply variations, Texas and Louisiana, USA, *in* Anderson, J.B., and Rodriguez, A.B., eds., Response of Gulf Coast Estuaries to Sea-Level Rise and Climate Change: Geological Society of America Special Paper 443, p. 65–88, https://doi.org/10.1130/2008.2443(05).

Milliken, K.T., Anderson, J.B., and Rodriguez, A.B., 2008c, Record of dramatic Holocene environmental changes linked to eustasy and climate change in Calcasieu Lake, Louisiana, USA, *in* Anderson, J.B., and Rodriguez, A.B., eds., Response of Gulf Coast Estuaries to Sea-Level Rise and Climate Change: Geological Society of America Special Paper 443, p. 43–63, https://doi.org/10.1130/2008.2443(04).

Milliken, K.T., Anderson, J.B., Simms, A., and Blum, M.D., 2017, A Holocene record of flux of alluvial sediment related to climate: Case studies from the Northern Gulf of Mexico: Journal of Sedimentary Research, v. 87, p. 780–794, https://doi.org/10.2110/jsr.2017.43.

Moore, L.J., List, J.H., Williams, S.J., and Stolper, D., 2010, Complexities in barrier island response to sea level rise: Insights from numerical model experiments, North Carolina Outer Banks: Journal of Geophysical Research: Earth Science, v. 115, F03004, https://doi.org/10.1029/2009JF001299.

Moore, S., Heise, E.A., Grove, M., Reisinger, A., and Benavides, J.A., 2021, Evaluating the impacts of dam construction and longshore transport upon modern sedimentation within the Rio Grande Delta (Texas, U.S.A.): Journal of Coastal Research, v. 37, p. 26–40.

Moran, K.E., Nittrouer, J.A., Perillo, M.M., Lorenzo-Trueba, J., and Anderson, J.B., 2017, Morphodynamic modeling of fluvial channel fill and avulsion time scales during early Holocene transgression, as substantiated by the incised valley stratigraphy of the Trinity River, Texas: Journal of Geophysical Research: Earth Surface, v. 122, p. 215–234, https://doi.org/10.1002/2015JF003778.

Morris, J.T., Sundareshwar, P.V., Nietch, C.T., Kjerfve, B., and Cahoon, D.R., 2002, Response of coastal wetlands to rising sea level: Ecology, v. 83, p. 2869–2877, https://doi.org/10.1890/0012-9658(2002)083[2869:ROCWTR]2.0.CO;2.

Morton, R.A., 1979, Temporal and spatial variations in shoreline changes and their implications, examples from the Texas Gulf Coast: Journal of Sedimentary Research, v. 49, p. 1101–1112.

Morton, R.A., 1994, Texas barriers, in Davis, R.A., ed., Geology of Holocene Barrier Islands: Berlin, Springer-Verlag, p. 75–114, https://doi.org/10.1007/978-3-642-78360-9_3.

Morton, R.A., and Pieper, M.J., 1975, Shoreline changes on Brazos Island and South Padre Island (Mansfield Channel to mouth of the Rio Grande): Austin, Texas, University of Texas at Austin, Bureau of Economic Geology Geological Circular 75-2, 39 p., https://doi.org/10.23867/GC7502D.

Morton, R.A., and Sallenger, A.H., 2003, Morphological impacts of extreme storms on sandy beaches and barriers: Journal of Coastal Research, v. 19, p. 560–573.

Morton, R.A., and Winker, J.R., 1979, Distribution and significance of coarse biogenic and clastic deposits on the Texas Inner Shelf: Transactions of the Gulf Coast Association of Geological Societies, v. 29, p. 352–364.

Morton, R.A., Gibeaut, J.C., and Paine, G., 1995, Meso-scale transfer of sand during and after storms: Implications for prediction of shoreline movement: Marine Geology, v. 126, p. 161–179, https://doi.org/10.1016/0025-3227(95)00071-6.

Morton, R.A., Miller, T.L., and Moore, L.J., 2004, National Assessment of Shoreline Change: Part 1 Historical Shoreline Changes and Associated Coastal Land Loss along the U.S. Gulf of Mexico: U.S. Geological Survey Open-File Report 2004-1043, 45 p.

Morton, R.A., Bernier, J.C., and Barras, J.A., 2006, Evidence of regional subsidence and associated interior wetland loss induced by hydrocarbon production, Gulf Coast region, USA: Environmental Geology, v. 50, p. 261–274, https://doi.org/10.1007/s00254-006-0207-3.

Murray, A.B., and Moore, L.J., 2018, Geometric constraints on long-term barrier migration: From simple to surprising, in Moore, L.J., and Murray, A.B., eds., Barrier Island Dynamics and Response to Changing Climate: Cham, Switzerland, Springer, p. 211–241, https://doi.org/10.1007/978-3-319-68086-6_7.

Nelson, H.F., and Bray, E.E., 1970, Stratigraphy and history of the Holocene sediments in the Sabine-High Island area, Gulf of Mexico, in Morgan, J.O., ed., Deltaic Sedimentation Modern and Ancient: Society of Economic Paleontologists and Mineralogists Special Publication 15, p. 48–77, https://doi.org/10.2110/pec.70.11.0048.

Nerem, R.S., Beckley, B.D., Fasullo, J.T., Hamlington, B.D., Masters, D., and Mitchum, G.T., 2018, Climate-change–driven accelerated sea-level rise detected in the altimeter era: Proceedings of the National Academy of Sciences of the United States of America, v. 115, p. 2022–2025, https://doi.org/10.1073/pnas.1717312115.

Nichol, S.L., Boyd, B., and Penland, S., 1996, Sequence stratigraphy of a coastal plain-incised estuary: Lake Calcasieu, southwest Louisiana: Journal of Sedimentary Research, v. 66, p. 847–857.

Nielsen-Gammon, J.W., Banner, J.L., Cook, B.I., Tremaine, D.M., Wong, C.I., Mace, R.E., Gao, H., Yang, Z.L., Gonzalez, M.F., Hoffpauir, R., and Gooch, T., 2020, Unprecedented drought challenges for Texas water resources in a changing climate: What do researchers and stakeholders need to know?: Earth's Future, v. 8, e2020EF001552, https://doi.org/10.1029/2020EF001552.

Nienhuis, J.H., and Lorenzo-Trueba, J., 2019, Can barrier islands survive sea-level rise? Quantifying the relative role of tidal inlets and overwash deposition: Geophysical Research Letters, v. 46, p. 14,613–14,621, https://doi.org/10.1029/2019GL085524.

Nordt, L.C., Boutton, T.W., Hallmark, C.T., and Waters, M.R., 1994, Late Quaternary vegetation and climate changes in central Texas based on the isotope composition of organic carbon: Quaternary Research, v. 41, p. 109–120, https://doi.org/10.1006/qres.1994.1012.

Nordt, L.C., Boutton, T.W., Jacob, J.S., and Mandel, R.D., 2002, C_4 plant productivity and climate-CO_2 variations in south-central Texas during the late Quaternary: Quaternary Research, v. 58, p. 182–188, https://doi.org/10.1006/qres.2002.2344.

Odezulu, C.I., Lorenzo-Trueba, J., Wallace, D.J., and Anderson, J.B., 2018, Follets Island: A case of unprecedented change and transition from rollover to subaqueous shoals, in Moore, L.J., and Murray, A.B., eds., Barrier Island Dynamics and Response to Changing Climate: Cham, Switzerland, Springer, p. 147–174, https://doi.org/10.1007/978-3-319-68086-6_5.

Odezulu, C.I., Swanson, T., and Anderson, J.B., 2020, Holocene progradation and retrogradation of the Central Texas Coast regulated by alongshore and cross-shore sediment flux variability: The Depositional Record: Journal of the International Association of Sedimentologists, v. 7, p. 77–92, https://doi.org/10.1002/dep2.130.

Oppenheimer, M., Glavovic, B.C., Hinkel, J., van de Wal, R., Magnan, A.K., Abd-Elgawad, A., Cai, R., Cifuentes-Jara, M., DeConto, R.M., Ghosh, T., Hay, J., Isla, F., Marzeion, B., Meyssignac, B., and Sebesvari, Z., 2019, Sea level rise and implications for low-lying islands, coasts and communities, in Pörtner, H.-O., Roberts, D.C., Masson-Delmotte, V., Zhai, P., Tignor, M.,. Poloczanska, E., Mintenbeck, K., Alegría, A., Nicolai, M., Okem, A., Petzold, J., Rama, B., and Weyer, N.M., eds., IPCC Special Report on the Ocean and Cryosphere in a Changing Climate: Cambridge, UK, Cambridge University Press, p. 321–445, https://doi.org/10.1017/9781009157964.006.

Otvos, E.G., 2001, Mississippi Coast: Stratigraphy and Quaternary Evolution in the Northern Gulf Coastal Plain Framework: U.S. Geological Survey Open-File Report 01-415-H, 59 p.

Otvos, E.G., 2005a, Coastal barriers, Gulf of Mexico: Holocene evolution and chronology: Journal of Coastal Research, v. 42, p. 141–163.

Otvos, E.G., 2005b, Numerical chronology of Pleistocene coastal plain and valley development; extensive aggradation during glacial low sea-levels: Quaternary International, v. 135, p. 91–113, https://doi.org/10.1016/j.quaint.2004.10.026.

Otvos, E.G., 2018, Coastal barriers, northern Gulf—Last eustatic cycle; genetic categories and development contrasts. A review: Quaternary Science Reviews, v. 193, p. 212–243, https://doi.org/10.1016/j.quascirev.2018.04.001.

Otvos, E.G., 2020, Coastal barriers—Fresh look at origins, nomenclature and classification issues: Geomorphology, v. 355, https://doi.org/10.1016/j.geomorph.2019.107000.

Otvos, E.G., and Howat, W.E., 1996, South Texas Ingleside Barrier; coastal sediment cycles and vertebrate fauna. Late Pleistocene stratigraphy revised: Transactions of the Gulf Coast Association of Geological Societies, v. 46, p. 333–344.

Paine, J.G., 1993, Subsidence of the Texas coast: Inferences from historical and late Pleistocene sea levels: Tectonophysics, v. 222, p. 445–458, https://doi.org/10.1016/0040-1951(93)90363-O.

Paine, J.G., Sojan, M., and Tiffany, C., 2012, Historical shoreline change through 2007, Texas Gulf Coast: Rates, contributing causes, and Holocene context: Gulf Coast Association of Geological Societies Journal, v. 1, p. 13–26.

Paine, J.G., Caudle, T.L., and Andrews, J.R., 2021, Shoreline Movement and Beach and Dune Volumetrics along the Texas Gulf Coast, 1930s to 2019: Texas General Land Office Final Report CEPRA Project 1662, 101 p.

Parker, R.H., 1959, Macro-invertebrate assemblages of central Texas coastal bays and Laguna Madre: American Association of Petroleum Geologists Bulletin, v. 43, p. 2100–2166.

Parker, R.H., 1960, Ecology and distributional patterns of marine macro-invertebrates, northern Gulf of Mexico, in Phleger, F.B., and Van Andel, T.H., eds., Recent Sediments, Northwest Gulf of Mexico: Tulsa, Oklahoma, American Association of Petroleum Geologists, p. 302–337.

Peltier, W.R., and Fairbanks, R.G., 2006, Global glacial ice volume and Last Glacial Maximum duration from an extended Barbados sea level record: Quaternary Science Reviews, v. 25, p. 3322–3337, https://doi.org/10.1016/j.quascirev.2006.04.010.

Phillips, J.D., 2010, Relative importance of intrinsic, extrinsic, and anthropic factors in the geomorphic zonation of the Trinity River, Texas: Journal of the American Water Resources Association, v. 46, p. 807–823, https://doi.org/10.1111/j.1093-474X.2010.00457.x.

Phillips, J.D., Slattery, M.C., and Musselman, Z.A., 2004, Dam-to-delta sediment inputs and storage in the lower Trinity River, Texas: Geomorphology, v. 62, p. 17–34, https://doi.org/10.1016/j.geomorph.2004.02.004.

Pilkey, O.H., and Davis, T.W., 1987, An analysis of coastal recession models: North Carolina coast, in Nummedal, D., Pilkey, O.H., and Howard, J., eds., Sea-Level Fluctuations and Coastal Evolution: Society for Sedimentary Geology (SEPM) Special Publication 41, p. 59–68, https://doi.org/10.2110/pec.87.41.0059.

Poag, C.W., 1981, Ecologic Atlas of Benthic Foraminifera of the Gulf of Mexico: Woods Hole, Massachusetts, Marine Science International, 174 p.

Price, W.A., 1933, Role of diastrophism in topography of Corpus Christi area, south Texas: American Association of Petroleum Geologists Bulletin, v. 17, p. 907–962.

Prothro, L.O., Majewski, W., Yokoyama, Y., Simkins, L.M., Anderson, J.B., Yamane, M., and Ohkouchi, N., 2020, The dynamic retreat of the East Antarctic Ice Sheet: An assessment of timing, pathways, and forcings: Quaternary Science Reviews, v. 230, https://doi.org/10.1016/j.quascirev.2020.106166.

Qu, F., Lu, Z., Zhang, Q., Bawden, G.W., Kim, J.W., Zhao, C., and Qu, W., 2015, Mapping ground deformation over Houston-Galveston, Texas using multi-temporal InSAR: Remote Sensing of Environment, v. 169, p. 290–306, https://doi.org/10.1016/j.rse.2015.08.027.

Raff, J., Shawler, J., Ciarletta, D.J., Hein, E.A., Lorenzo-Trueba, J., and Hein, C., 2018, Insights into barrier-island stability derived from transgressive/regressive state changes of Parramore Island, Virginia: Marine Geology, v. 403, p. 1–19, https://doi.org/10.1016/j.margeo.2018.04.007.

Rasmussen, S.O., Andersen, K.K., Svensson, A.M., Steffensen, J.P., Vinther, B.M., Clausen, H.B., Siggaard-Andersen, M.L., Johnsen, S.J., Larsen, L.B., Dahl-Jensen, D., and Bigler, M., 2006, A new Greenland ice core chronology for the last glacial termination: Journal of Geophysical Research: Atmospheres, v. 111, D06102, https://doi.org/10.1029/2005JD006079.

Ravens, T.M., Thomas, R.C., Roberts, K.A., and Santschi, P.H., 2009, Causes of salt marsh erosion in Galveston Bay, Texas: Journal of Coastal Research, v. 25, p. 265–272.

Reeves, I.R.B., Moore, L.J., Murray, A.B., Anarde, K.A., and Goldstein, E.B., 2021, Dune dynamics drive discontinuous barrier retreat: Geophysical Research Letters, v. 48, e2021GL092958, https://doi.org/10.1029/2021GL092958.

Rehkemper, J.L., 1969, Sedimentology of Holocene estuarine deposits, Galveston Bay, in Lankford, R.R., and Rogers, J.J.W., eds., Holocene Geology of the Galveston Bay Area: Houston, Texas, Houston Geological Society, p. 12–52.

Reimer, P.J., Bard, E., Bayliss, A., Beck, J.W., Blackwell, P.G., Ramsey, C.B., Buck, C.E., Cheng, H., Edwards, R.L., Friedrich, M., Grootes, P.M., Guilderson, T.P., Haflidason, H., Hajdas, I., Hatté, C., Heaton, T.J., Hoffmann, D.L., Hogg, A.G., Hughen, K.A., Kaiser, K.F., Kromer, B., Manning, S.W., Niu, M., Reimer, R.W., Richards, D.A., Scott, E.M., Southon, J.R., Staff, R.A., Turney, C.S.M., and van der Plicht J., 2013, IntCal13 and Marine13 radiocarbon age calibration curves 0–50,000 years cal BP: Radiocarbon, v. 55, p. 1869–1887, https://doi.org/10.2458/azu_js_rc.55.16947.

Rice, J.A., Simms, A.R., Buzas-Stephens, P., Steel, E., Livsey, D., Reynolds, L.C., Yokoyama, Y., and Halihan, T., 2020, Deltaic response to climate change: The Holocene history of the Nueces Delta: Global and Planetary Change, v. 191, https://doi.org/10.1016/j.gloplacha.2020.103213.

Rignot, E., 2021, Sea level rise from melting glaciers and ice sheets caused by climate warming above pre-industrial levels: Physics-Uspekhi, https://doi.org/10.3367/UFNe.2021.11.039106

Rodriguez, A.B., Anderson, J.B., Siringan, F.P., and Taviani, M., 1999, Sedimentary facies and genesis of Holocene sand banks on the east Texas inner continental shelf, in Sneddin, J., and Bergman, K., eds., Isolated Shallow Marine Sand Bodies: Society for Sedimentary Geology (SEPM) Special Publication 64, p. 165–178, https://doi.org/10.2110/pec.99.64.0165.

Rodriguez, A.B., Hamilton, M., and Anderson, J.B., 2000, Facies and evolution of the modern Brazos Delta, Texas: Wave versus flood influence: Journal of Sedimentary Research, v. 70, p. 283–295, https://doi.org/10.1306/2DC40911-0E47-11D7-8643000102C1865D.

Rodriguez, A.B., Fassell, M., and Anderson, J.B., 2001, Variations in shoreface progradation and ravinement along the Texas coast, Gulf of Mexico: Sedimentology, v. 48, p. 837–853, https://doi.org/10.1046/j.1365-3091.2001.00390.x.

Rodriguez, A.B., Anderson, J.B., Siringan, F.P., and Taviani, M., 2004, Holocene evolution of the east Texas Coast and inner continental shelf: Along-strike variability in coastal retreat rates: Journal of Sedimentary Research, v. 74, p. 404–421, https://doi.org/10.1306/092403740405.

Rodriguez, A.B., Anderson, J.B., and Simms, A.R., 2005, Terrace inundation as an autocyclic mechanism for parasequence formation: Galveston estuary, Texas: Journal of Sedimentary Research, v. 75, p. 608–620, https://doi.org/10.2110/jsr.2005.050.

Rodriguez, A.B., Simms, A., and Anderson, J., 2010, Bay-head deltas across the northern Gulf of Mexico back step in response to the 8.2 ka cooling event: Quaternary Science Reviews, v. 29, p. 3983–3993, https://doi.org/10.1016/j.quascirev.2010.10.004.

Rodriguez, A.B., Yu, W., and Theuerkauf, E.J., 2018, Abrupt increase in washover deposition along a transgressive barrier island during the late nineteenth century acceleration of sea-level rise, in Moore, L.J., and Murray, A.B., eds., Barrier Island Dynamics and Response to Changing Climate: Cham, Switzerland, Springer, p. 121–145, https://doi.org/10.1007/978-3-319-68086-6_4.

Rodriguez, A.B., McKee, B.A., Miller, C.B., Bost, M.C., and Atencio, A.N., 2020, Coastal sedimentation across North America doubled in the 20th century despite river dams: Nature Communications, v. 11, 3249, https://doi.org/10.1038/s41467-020-16994-z.

Rodgers, N., and Paola, C., 2021, Intermittent retreat behavior in experimental barrier island response to constant sea level rise and wave forcing: Journal of Geophysical Research–Earth Surface, v. 126, e2021JF006086, https://doi.org/10.1029/2021JF006086.

Russell, B.T., Risser, M.D., Smith, R.L., and Kunkel, K.E., 2020, Investigating the association between late spring Gulf of Mexico sea surface temperatures and US Gulf Coast precipitation extremes with focus on Hurricane Harvey: Environmetrics, v. 31, https://doi.org/10.1002/env.2595.

Russell, R.J., and Howe, H.V., 1935, Cheniers of southwestern Louisiana: Geographical Review, v. 25, p. 449–461, https://doi.org/10.2307/209313.

Shackleton, N.J., 1987, Oxygen isotopes, ice volume and sea level: Quaternary Science Reviews, v. 6, p. 183–190, https://doi.org/10.1016/0277-3791(87)90003-5.

Shawler, J.L., Ciarletta, D.J., Connell, J.E., Boggs, B.Q., Lorenzo-Trueba, J., and Hein, C.J., 2020, Relative influence of antecedent topography and sea-level rise on barrier-island migration: Sedimentology, v. 68, p. 639–669, https://doi.org/10.1111/sed.12798.

Shen, Z., Dawers, N.H., Törnqvist, T.E., Gasparini, N.M., Hijma, M.P., and Mauz, B., 2017, Mechanisms of late Quaternary fault throw-rate variability along the north central Gulf of Mexico coast: Implications for coastal subsidence: Basin Research, v. 29, p. 557–570, https://doi.org/10.1111/bre.12184.

Shepard, F.P., and Moore, D.G., 1955, Central Texas coast sedimentation: Characteristics of sedimentary environment, recent history, and diagenesis: Part 2: American Association of Petroleum Geologists Bulletin, v. 39, p. 1463–1593, https://archives.datapages.com/data/bulletns/1953-56/data/pg/0039/0008/1500/1463.htm.

Shideler, G.L., 1978, A sediment-dispersal model for the South Texas continental shelf, northwest Gulf of Mexico: Marine Geology, v. 26, p. 289–313, https://doi.org/10.1016/0025-3227(78)90064-6.

Shideler, G.L., 1986, Seismic and physical stratigraphy of late Quaternary deposits, south Texas coastal complex, in Shideler, G.L., ed., Stratigraphic Studies of a Late Quaternary Barrier-Type Coastal Complex: Mustang Island–Corpus Christi Area, South Texas Gulf Coast: U.S. Geological Survey Professional Paper 1328-B, p. 9–31.

Siegert, M., Alley, R.B., Rignot, E., Englander, J., and Corell, R., 2020, Twenty-first century sea-level rise could exceed IPCC projections for strong-warming futures: One Earth, v. 3, p. 691–703, https://doi.org/10.1016/j.oneear.2020.11.002.

Simms, A.R., 2021, Last Interglacial Sea Levels within the Gulf of Mexico: Earth System Science Data Discussions, v. 13, p. 1419–1439, https://doi.org/10.5194/essd-13-1419-2021.

Simms, A.R., and Rodriguez, A.B., 2014, Where do coastlines stabilize following rapid retreat?: Geophysical Research Letters, v. 41, p. 1698–1703, https://doi.org/10.1002/2013GL058984.

Simms, A.R., and Rodriguez, A.B., 2015, The influence of valley morphology on the rate of bayhead delta progradation: Journal of Sedimentary Research, v. 85, p. 38–44, https://doi.org/10.2110/jsr.2015.02.

Simms, A.R., Anderson, J.B., and Blum, M., 2006, Mustang Island, an example of an aggradational barrier island: Sedimentary Geology, v. 187, p. 105–125, https://doi.org/10.1016/j.sedgeo.2005.12.023.

Simms, A.R., Lambeck, K., Purcell, A., Anderson, J., and Rodriguez, A., 2007a, Sea-level history for the Gulf of Mexico since the Last Glacial Maximum with implications for the melting of the Laurentide Ice Sheet: Quaternary Science Reviews, v. 26, p. 920–940, https://doi.org/10.1016/j.quascirev.2007.01.001.

Simms, A.R., Anderson, J.B., Milliken, K.T., Taha, Z.P., and Wellner, J.S., 2007b, Geomorphology and age of the oxygen isotope stage 2 (last lowstand) sequence boundary on the northwestern Gulf of Mexico continental shelf, in Davies, R.J., Posamentier, H.W., Wood, L.J., and Cartwright, J.A., eds., Seismic Geomorphology: Applications to Hydrocarbon Exploration and Production: Geological Society, London, Special Publication 277, p. 29–46, https://doi.org/10.1144/GSL.SP.2007.277.01.03.

Simms, A.R., Anderson, J.B., Rodriguez, A.B., and Taviani, M., 2008, Mechanisms controlling environmental change within an estuary: Corpus Christi Bay, Texas, USA, in Anderson, J.B., and Rodriguez, A.B., eds., Response of Gulf Coast Estuaries to Sea-Level Rise and Climate Change:

Geological Society of America Special Paper 443, p. 121–146, https://doi.org/10.1130/2008.2443(08).

Simms, A.R., Aryal, N., Yokoyama, Y., Matsuzaki, H., and DeWitt, R., 2009, Insights on a proposed mid-Holocene highstand along the northwestern Gulf of Mexico from the evolution of small coastal ponds: Journal of Sedimentary Research, v. 79, p. 757–772, https://doi.org/10.2110/jsr.2009.079.

Simms, A.R., Aryal, N., Miller, L., and Yokoyama, Y., 2010, The incised valley of Baffin Bay, Texas: A tale of two climates: Sedimentology, v. 57, p. 642–669, https://doi.org/10.1111/j.1365-3091.2009.01111.x.

Simms, A.R., Anderson, J.B., DeWitt, R., Lambeck, K., and Purcell, A., 2013, Calculating rates of coastal subsidence from the last interglacial (LIG) shorelines: A late Pleistocene Gulf of Mexico perspective illustrating the importance of sediment loading: Global and Planetary Change, v. 111, p. 296–308, https://doi.org/10.1016/j.gloplacha.2013.10.002.

Sionneau, T., Bout-Roumazeilles, V., Biscaye, P.E., Van Vliet-Lanoe, B., and Bory, A., 2008, Clay mineral distributions in and around the Mississippi River watershed and Northern Gulf of Mexico: Sources and transport patterns: Quaternary Science Reviews, v. 27, p. 1740–1751, https://doi.org/10.1016/j.quascirev.2008.07.001.

Siringan, F.P., and Anderson, J.B., 1993, Seismic facies, architecture, and evolution of the Bolivar Roads tidal inlet/delta complex, East Texas Gulf Coast: Journal of Sedimentary Petrology, v. 63, p. 794–808.

Siringan, F.P., and Anderson, J.B., 1994, Modern shoreface and inner-shelf storm deposits off the East Texas Coast, Gulf of Mexico: Journal of Sedimentary Research, v. B64, p. 99–110.

Smyth, W.C., Anderson, J.B., and Thomas, M.A., 1988, Seismic facies analysis on entrenched valley-fill: A case study in Galveston Bay, Texas: Transactions of the Gulf Coast Association of Geological Societies, v. 38, p. 385–394.

Snedden, J.W., Nummedal, D., and Amos, A.F., 1988, Storm and fairweather combined flow on the central Texas continental shelf: Journal of Sedimentary Petrology, v. 58, p. 580–595.

Snow, J.N., 1998, Late Quaternary highstand and transgressive deltas of the ancestral Colorado River: Eustatic and climate controls on deposition [M.A. thesis]: Houston, Texas, Rice University, 138 p.

Stolper, D., List, J.H., and Thieler, E.R., 2005, Simulating the evolution of coastal morphology and stratigraphy with a new morphological behavior model (GEOMBEST): Marine Geology, v. 218, p. 17–36, https://doi.org/10.1016/j.margeo.2005.02.019.

Suter, J.R., and Berryhill, H.L., 1985, Late Quaternary shelf-margin deltas: Northwest Gulf of Mexico: American Association of Petroleum Geologists Bulletin, v. 69, p. 77–91.

Swift, D.J.P., 1968, Coastal erosion and transgressive stratigraphy: The Journal of Geology, v. 76, p. 444–456, https://doi.org/10.1086/627342.

Swift, D.J.P., Niederoda, A.W., Vincent, C.E., and Hopkins, T.S., 1985, Barrier island evolution, middle Atlantic shelf, USA, Part I: Shoreface dynamics: Marine Geology, v. 63, p. 331–361, https://doi.org/10.1016/0025-3227(85)90089-1.

Sylvia, D.A., and Galloway, W.E., 2006, Morphology and stratigraphy of the late Quaternary lower Brazos valley: Implications for paleo-climate, discharge and sediment delivery: Sedimentary Geology, v. 190, p. 159–175, https://doi.org/10.1016/j.sedgeo.2006.05.023.

Syvitski, J.P., and Milliman, J.D., 2007, Geology, geography, and humans battle for dominance over the delivery of fluvial sediment to the coastal ocean: The Journal of Geology, v. 115, p. 1–19, https://doi.org/10.1086/509246.

Taha, P.Z., and Anderson, J.B., 2008, The influence of valley aggradation and listric normal faulting on styles of river avulsion: A case study of the Brazos River, Texas, USA: Geomorphology, v. 95, p. 429–448, https://doi.org/10.1016/j.geomorph.2007.07.014.

Thomas, M.A., and Anderson, J.B., 1989, Glacial eustatic controls on seismic sequences and parasequences of the Trinity/Sabine incised valley, Texas Continental Shelf: Transactions of the Gulf Coast Association of Geological Societies, v. 39, p. 563–570.

Thomas, M.A., and Anderson, J.B., 1994, Sea-level controls on the facies architecture of the Trinity/Sabine incised-valley system, Texas continental shelf, in Dalrymple, R., Boyd, R., and Zaitlin, B.A., eds., Incised Valley Systems: Origin and Sedimentary Sequences: Society for Sedimentary Geology (SEPM) Special Publication 51, p. 63–82, https://doi.org/10.2110/pec.94.12.0063.

Toomey, R.S., Blum, M.D., and Valastro, S., 1993, Late Quaternary climates and environments of the Edwards Plateau, Texas: Global and Planetary Change, v. 7, p. 299–320, https://doi.org/10.1016/0921-8181(93)90003-7.

Törnqvist, T.E., and Hijma, M.P., 2012, Links between early Holocene ice-sheet decay, sea-level rise and abrupt climate change: Nature Geoscience, v. 5, p. 601–606, https://doi.org/10.1038/ngeo1536.

Törnqvist, T.E., González, J.L., Newsom, L.A., Van der Borg, K., De Jong, A.F.M., and Kurnik, C.W., 2004a, Deciphering Holocene sea-level history on the U.S. Gulf Coast: A high resolution record from the Mississippi Delta: Geological Society of America Bulletin, v. 116, p. 1026–1039, https://doi.org/10.1130/B2525478.1.

Törnqvist, T.E., Bick, S.J., González, J.L., Van der Borg, K., and De Jong, A.F.M., 2004b, Tracking the sea-level signature of the 8.2 ka cooling event: New constraints from the Mississippi Delta: Geophysical Research Letters, v. 31, L23309, https://doi.org/10.1029/2004GL021429.

Törnqvist, T.E., Bick, S.J., Van der Borg, K., and De Jong, A.F.M., 2006, How stable is the Mississippi Delta?: Geology, v. 34, p. 697–700, https://doi.org/10.1130/G22624.1.

Törnqvist, T.E., Wallace, D.J., Storms, J.E.A., Wallinga, J., van Dam, R.L., Blaauw, M., Derksen, M.S., Klerks, C.J.W., Meijneken, C., and Snijders, E.M.A., 2008, Mississippi Delta subsidence primarily caused by compaction of Holocene strata: Nature Geoscience, v. 1, p. 173–176, https://doi.org/10.1038/ngeo129.

Törnqvist, T.E., Jankowski, K.L., Li, Y-X, and Gonzalez, J.L., 2020, Tipping points of Mississippi Delta marshes due to accelerated sea-level rise: Science Advances, v. 6. eaaz5512, https://doi.org/10.1126/sciadv.aaz5512.

Toscano, M.A., and Macintyre, I.G., 2003, Corrected western Atlantic sea-level curve for the last 11,000 years based on calibrated ^{14}C dates from Acropora palmata framework and intertidal mangrove peat: Coral Reefs, v. 22, p. 257–270, https://doi.org/10.1007/s00338-003-0315-4.

Troiani, B.T., Simms, A.R., Dellapenna, T., Piper, E., and Yokoyama, Y., 2011, The importance of sea-level and climate change, including changing wind energy, on the evolution of a coastal estuary: Copano Bay, Texas: Marine Geology, v. 280, p. 1–19, https://doi.org/10.1016/j.margeo.2010.10.003.

Van Heijst, W.I.M., Postma, G., Meijer, X.D., Snow, J.N., and Anderson, J.B., 2001, Quantitative analogue flume-model study of the Late Quaternary Colorado River-Delta evolution: Basin Research, v. 13, p. 243–268, https://doi.org/10.1046/j.1365-2117.2001.00150.x.

Verbeek, E.R., 1979, Surface faults in the Gulf Coastal Plain between surface faults in the Gulf Coastal Plain between Victoria and Beaumont, Texas: Tectonics, v. 52, p. 373–375.

Villarini, G., and Vecchi, G.A., 2013, Projected increases in North Atlantic tropical cyclone intensity from CMIP5 models: Journal of Climatology, v. 26, p. 3231–3240, https://doi.org/10.1175/JCLI-D-12-00441.1.

Vinther, B.M., Clausen, H.B., Johnsen, S.J., Rasmussen, S.O., Andersen, K.K., Buchardt, S.L., Dahl-Jensen, D., Seierstad, I.K., Siggaard-Andersen, M.L., Steffensen, J.P., and Svensson, A., 2006, A synchronized dating of three Greenland ice cores throughout the Holocene: Journal of Geophysical Research: Atmospheres, v. 111, D13102, https://doi.org/10.1029/2005JD006921.

Wallace, D.J., and Anderson, J.B., 2010, Evidence of similar frequency intense hurricane strikes for the Gulf of Mexico over the late Holocene: Geology, v. 38, p. 511–514, https://doi.org/10.1130/G30729.1.

Wallace, D.J., and Anderson, J.B., 2013, Unprecedented erosion of the upper Texas coast: Response to accelerated sea-level rise and hurricane impacts: Geological Society of America Bulletin, v. 125, p. 728–740, https://doi.org/10.1130/B30725.1.

Wallace, D.J., and Woodruff, J.D., 2020, Experimental and numerical models of fine sediment transport by tsunamis, in Engel, M., Pilarczyk, J., May, S.M., Brill, D., and Garrett, E., eds., Geological Records of Tsunamis and Other Extreme Waves: Amsterdam, Elsevier, p. 491–509, https://doi.org/10.1016/B978-0-12-815686-5.00023-7.

Wallace, D.J., Anderson, J.B., and Rodriguez, A.B., 2009, Natural versus anthropogenic mechanisms of erosion along the upper Texas coast, in Kelley, J.T., Pilkey, O.H., and Cooper, J.A.G., eds., America's Most Vulnerable Coastal Communities: Geological Society of America Special Paper 460, p. 137–147, https://doi.org/10.1130/2009.2460(10).

Wallace, D.J., Anderson, J.B., and Fernández, R.A., 2010, Transgressive ravinement versus depth of closure: A geological perspective from the upper Texas coast: Journal of Coastal Research, v. 26, p. 1057–1067, https://doi.org/10.2112/JCOASTRES-D-10-00034.1.

Wallace, D.J., Woodruff, J.D., Anderson, J.B., and Donnelly, J.P., 2014, Palaeo-hurricane reconstructions from sedimentary archives along the Gulf of Mexico, Caribbean Sea, and western North Atlantic Ocean margins, in Martini, I.P., and Wanless, H.R., eds., Sedimentary Coastal Zones from

High to Low Latitudes: Similarities and Differences: Geological Society, London, Special Publication 388, p. 481–501, https://doi.org/10.1144/SP388.12

Wang, J., Church, J.A., Zhang, X., and Chen, X., 2021, Reconciling global mean and regional sea level change in projections and observations: Nature Communications, v. 12, 990, https://doi.org/10.1038/s41467-021-21265-6.

Wantland, K.F., 1969, Distribution of modern brackish-water foraminifera in Trinity Bay, *in* Lankford, R.R., and Rogers, J.J.W., eds., Holocene Geology of the Galveston Bay Area: Houston, Texas, Houston Geological Society, p. 93–118.

Weight, R., Anderson, J.B., and Fernandez, R., 2011, Rapid mud accumulation on the central Texas shelf linked to climate change and sea-level rise: Journal of Sedimentary Research, v. 81, p. 743–764, https://doi.org/10.2110/jsr.2011.57.

Wellner, J.S., Sarzalejo, S., Logoe, M., and Anderson, J.B., 2004, The Late Quaternary stratigraphic evolution of the west Louisiana/East Texas continental shelf, Late Quaternary stratigraphic evolution of the northern Gulf of Mexico: A synthesis, *in* Anderson, J.B., and Fillon, R.H., eds., Late Quaternary Stratigraphic Evolution of the Northern Gulf of Mexico Basin: Society for Sedimentary Geology (SEPM) Special Publication 79, p. 217–236, https://doi.org/10.2110/pec.04.79.0217.

Wells, J.T., 1983, Dynamics of coastal fluid muds in low-, moderate-, and high-tide-range environments: Canadian Journal of Fisheries and Aquatic Sciences, v. 40, no. S1, p. s130–s142, https://doi.org/10.1139/f83-276.

Wells, J.T., and Coleman, J.M., 1981, Physical processes and fine-grained sediment dynamics, coast of Surinam, South America: Journal of Sedimentary Research, v. 51, p. 1053–1068, https://doi.org/10.1306/212F7E1E-2B24-11D7-8648000102C1865D.

White, W.A., Calnan, T.R., Morton, R.A., Kimble, R.S., Littleton, T.G., McGowen, J.H., Nance, H.S., and Schmedes, K.E., 1983, Submerged Lands of Texas, Corpus Christi Area: Sediments, Geochemistry, Benthic Macroinvertebrates, and Associated Wetlands: Austin, Texas, University of Texas at Austin, Bureau of Economic Geology, 154 p.

White, W.A., Morton, R.A., and Holmes, C.W., 2002, A comparison of factors controlling sedimentation rates and wetland loss in fluvial–deltaic systems, Texas Gulf Coast: Geomorphology, v. 44, p. 47–66, https://doi.org/10.1016/S0169-555X(01)00140-4.

Wilkinson, B.H., 1975, Matagorda Island, Texas: The evolution of a Gulf Coast barrier complex: Geological Society of America Bulletin, v. 86, p. 959–967, https://doi.org/10.1130/0016-7606(1975)86<959:MITTEO>2.0.CO;2.

Wilkinson, B.H., and Basse, R.A., 1978, Late Holocene history of the central Texas coast from Galveston Island to Pass Cavallo: Geological Society of America Bulletin, v. 89, p. 1592–1600, https://doi.org/10.1130/0016-7606(1978)89<1592:LHHOTC>2.0.CO;2.

Williams, S.J., Prins, D.A., and Meisburger, E.P., 1979, Sediment Distribution, Sand Resources and Geologic Character of the Inner Continental Shelf off Galveston County, Texas: U.S. Army Corps of Engineers Miscellaneous Report 79–4, 159 p.

Wong, C.I., Banner, J.L., and Musgrove, M., 2015, Holocene climate variability in Texas, USA: An integration of existing paleoclimate data and modeling with a new, high-resolution speleothem record: Quaternary Science Reviews, v. 127, p. 155–173, https://doi.org/10.1016/j.quascirev.2015.06.023.

Worrall, D.M., 2021, The Prehistory of Houston and Southeast Texas: Fulshear, Texas, Concertina Press, 485 p.

Wu, W., Biber, P., and Bethel, M., 2017, Thresholds of sea-level rise rate and sea-level rise acceleration rate in a vulnerable coastal wetland: Ecology and Evolution, v. 7, p. 10,890–10,903, https://doi.org/10.1002/ece3.3550.

Wu, W., Biber, P., Mishra, D.R., and Ghosh, S., 2020, Sea-level rise thresholds for stability of salt marshes in a riverine versus a marine dominated estuary: The Science of the Total Environment, v. 718, https://doi.org/10.1016/j.scitotenv.2020.137181.

Yeager, K.M., Wolfe, P.C., Feagin, R.A., Brunner, C.A., and Schindler, K.J., 2019, Active near-surface growth faulting and late Holocene history of motion: Matagorda Peninsula, Texas: Geomorphology, v. 327, p. 159–169, https://doi.org/10.1016/j.geomorph.2018.10.019.

Yokoyama, Y., Lambeck, K., De Deckker, P., Johnston, P., and Fifield, L.K., 2000, Timing of the Last Glacial Maximum from observed sea-level minima: Nature, v. 406, p. 713–716, https://doi.org/10.1038/35021035.

Yu, S.-Y., Berglund, B.E., Sandgren, P., and Lambeck, K., 2007, Evidence for a rapid sea-level rise 7600 yr ago: Geology, v. 35, p. 891–894, https://doi.org/10.1130/G23859A.1.

Yuill, B., Lavoie, D., and Reed, D.J., 2009, Understanding subsidence processes in coastal Louisiana: Journal of Coastal Research, 10054, p. 23–36, https://doi.org/10.2112/SI54-012.1.

Zhang, W., Villarini, G., Vecchi, G.A., and Smith, J.A., 2018, Urbanization exacerbated the rainfall and flooding caused by hurricane Harvey in Houston: Nature, v. 563, p. 384–388, https://doi.org/10.1038/s41586-018-0676-z.

Zhu, L., Quiring, S.M., Guneralp, I., and Peacock, W.G., 2015, Variations in tropical cyclone-related discharge in four watersheds near Houston, Texas: Climate Risk Management, v. 7, p. 1–10, https://doi.org/10.1016/j.crm.2015.01.002.

MANUSCRIPT ACCEPTED BY THE SOCIETY 10 JANUARY 2022
MANUSCRIPT PUBLISHED ONLINE 7 JUNE 2022